Praise for *Cyber Crisis*

"Strong cybersecurity is essential for every individual and business in this time of elevated threats. In *Cyber Crisis*, Dr. Cole provides cutting-edge, real-world advice on how to protect your business and your family from today's persistent cyber threats."

—Andrew McCabe, #1 *New York Times* bestselling author of *The Threat* and former deputy FBI director

"The more I work with high-profile individuals, I realize the impact that cybersecurity can have on their lives. Anyone and everyone has to pay attention to cybersecurity and there is no one better than Dr. Cole. When my friends and family need help or advice in the area of cybersecurity, Dr. Cole is the one I call and he has personally helped me on several occasions. He is the best in the business. With *Cyber Crisis*, I now have a resource that I can recommend to them."

—Tim Storey, celebrity life coach

"Eric Cole is my 'go to' authority on cybersecurity. Not only is he an expert, he's an expert explainer, which is invaluable to both businesses and the media. *Cyber Crisis* does a top-notch job of explaining cybersecurity in a way that anyone can understand. If you want your company or your audience to stay ahead of the hacks, call Eric and read his book. I recommend him without reservation."

—Joel Roberts, former host for KABC Radio, Los Angeles

"*Cyber Crisis* is a book for anyone and everyone. When first asked to read the book, I was thinking 'What do I know or need to know about cyber security?' Well, I was wrong, dead wrong. As we would say in the NYPD, this book is the real deal. An easy read and unbelievably informative and eye opening. Whether you are a parent, business owner, CEO, CFO, governmental official, or an everyday hardworking individual that uses a smartphone or computer, you will learn something and not regret reading this book. Those who are company leaders and cybersecurity personnel, this is the cliff notes version on how to keep your job. Having personally known Dr. Eric Cole and his beautiful family for over ten years, I was already aware of his experience, background, and work ethic. It is astonishingly impressive, and after the first chapter you will know why I think so. He is the absolute best in the world at what he does. There is a reason that many of the most powerful and affluent people of the world have Eric's personal cell phone number. After you read this book, you will understand the reason."

—Peter Clark, NYPD Lieutenant Commander-Detective Squad (Retired)

"Cybersecurity is one of the top threats facing any business or organization. In *Cyber Crisis*, Dr. Cole emphasizes and concisely articulates the importance to every executive of prioritizing this critical threat. As a former military and civilian executive, I have broad-based experience in sensitive, data-centric environments and a great appreciation of the value of leadership awareness of and preparation for cybersecurity. Dr. Cole does a superb job of taking a very complex topic and presenting it in a way that is comprehensible and actionable for any type of industry or business. This book is an essential read for every executive in any industry."

—Jim Finkelstein, Rear Admiral for the US Navy (Retired)

"As a named partner of a major international law firm and former CEO of an international Fortune 150 company, I have repeatedly served in executive leadership positions of data-driven organizations. Cybersecurity is frequently at the forefront of strategy and investment planning and so often the leadership responsible for securing sensitive data has only a superficial understanding of the elements of true cybersecurity. Dr. Cole addresses this problem head-on in his new book, *Cyber Crisis*. This book, unlike any I have seen in my career, presents critical issues in a concise and easy to follow manner that most anyone can understand. This is truly required reading for all executives and leaders."

—Marshall Manley, former president and CEO of City Investing
Company and chairman of Home Insurance Company

"Cybersecurity is at the forefront of the medical industry. With numerous confidentiality and compliance requirements and continuously evolving data-sharing needs, the importance of awareness and preparation around data security has never been greater. From hospital executives to practitioners to third-party payers, the responsibility for data security is pervasive. *Cyber Crisis* by Dr. Cole plain and simply equips leaders with a working knowledge of cybersecurity and guides them concisely on how to prepare for and manage security threats. This book simplifies a challenging and crucial topic for our industry. It should be a staple read in the medical and associated industries."

—Paul M. Zimmerman, MD, founder of Automated
Healthcare Solutions and chairman of Gensco Pharma

"Leading an international distribution and supply company, we face difficult and growing data security challenges daily. Leaders are thrust into positions of responsibility for vast amounts of data management, often with inadequate baseline knowledge of cybersecurity practices and awareness. *Cyber Crisis* hits this issue head-on, offering a brilliant presentation of a complex topic in a methodical, consumable format that enables non-technical leadership to rapidly grasp and prepare for cyber threats. This book will be the gold standard for preparing senior leadership to manage this exploding threat."

—William Costlow, president of Performance Marketing

"Cybersecurity is one of the top threats facing any business. In *Cyber Crisis*, Dr. Cole emphasizes the importance of not ignoring this critical threat and making it a top priority. Dr. Cole does a great job of taking a very complex topic and making it easy to understand for any business. This book is a must-read for any executive in any business vertical."

—Amit Yoran, chairman and CEO of Tenable and former CEO of RSA

"As a retired Naval Officer who has worked routinely in the world of cyber threats, I have seen the disastrous consequences of data breaches, data compromises, and denial of service attacks and how destructive they can be to an enterprise. Dr. Cole's brilliant book emphasizes the importance of personal and institutional focus on this critical threat and making it a top priority at work and at home. Often cyber discussions appear too academic and hard to understand. You will find *Cyber Crisis* to be a reader-friendly primer on every aspect of cyber threats and should be considered a must-read for any business vertical."

—Edward "Sonny" Masso, Rear Admiral for the US Navy (Retired)—Flagship Connection

CYBER CRISIS

Protecting Your Business
from Real Threats
in the Virtual World

ERIC COLE

BenBella Books, Inc.
Dallas, TX

BenBella Books, Inc.
10440 N. Central Expressway
Suite 800
Dallas, TX 75231
benbellabooks.com
Send feedback to feedback@benbellabooks.com

BenBella is a federally registered trademark.

Printed in the United States of America
10 9 8 7 6 5 4 3 2 1

Library of Congress Control Number: 2020051871
ISBN 9781950665839 (trade cloth)
ISBN 9781953295279 (ebook)

Editing by Scott Calamar
Proofreading by Denise Pangia and Cape Cod Compositors, Inc.
Indexing by Debra Bowman
Text design and composition by PerfecType, Nashville, TN
Cover design by Pete Garceau
Printed by Lake Book Manufacturing

Special discounts for bulk sales are available. Please contact
bulkorders@benbellabooks.com.

To my amazing family, friends, and colleagues who always believe in me and support me in my mission to make cyberspace a safe place to live, work, and raise a family.

To everyone who is reading this book and taking the first step to increase their knowledge of cybersecurity. Whether we realize it or not, everyone and every business is a target, and to be safe in cyberspace, it is critical to embrace the fact that cybersecurity is your responsibility. I acknowledge you for taking this first step, and Godspeed on your journey.

Contents

Introduction | *The Current Reality* | 1

ONE | We Are All Targets | 11

TWO | We Live in Cyberspace | 37

THREE | The Hackers Are Here | 67

FOUR | Mobile Weaknesses | 87

FIVE | Your Life, Hanging in the Cloud | 105

SIX | They're in Your Business | 127

SEVEN | National Infrastructure Attack | 157

EIGHT | Cyberspace: A Place with No Borders | 201

NINE | Surviving the Cyber Crisis | 213

Conclusion | *Ten Lessons to Remember* | 223

Index | 237

About the Author | 245

Introduction

The Current Reality

If you watch the news, you might think there's a major cybersecurity threat every four to five months. The reality is that there's a cybersecurity attack every minute of every day. As one of the country's leading experts in cybersecurity, I often appear on the news as an authority when they decide to cover an attack. On the flip side, when I hear about new breaches not yet reported, I'll often call up the major media outlets and tell them.

"Hey, there's a breach that's going to come out that I want to give you the exclusive on, where forty-five million records were stolen."

News directors and reporters don't even flinch.

"Eric, unless it's over a hundred million, nobody cares," they say. "It's not newsworthy because it's just happening with such frequency."

There are at least one to two breaches where over one hundred million people's information is hacked and compromised every month. For breaches in the millions, we're talking daily.

Even when the news does cover cybersecurity breaches, they often don't cover them properly because they don't understand them. When the Marriott breach happened in 2018—one of the largest—half a billion records were stolen. It was on and off the news in less than forty-eight hours. Two days later, you couldn't even find any information about it anywhere. During the impeachment coverage of President Trump, there was an attack in which 320 million records were compromised, but no one covered it because the media did not view it as a high priority. If you look at other major news stories, they are out there for weeks, months, and sometimes years. So, unfortunately, if you are relying on the news to understand the cybersecurity landscape, you're getting a false impression that cybersecurity isn't as bad as you think. You're not even getting close to the full story of how dangerous the problem really is.

Breaches are happening to individuals and businesses on a regular basis. They're happening right now as you read this—and no one is talking about it.

Information has value. Whether you're a professional in a very small business or a very successful entrepreneur, I can guarantee you several things. First, you have an identity. Second, you, your business, or the business you work for has some critical data or critical information. And third, you have some money, even if it's only a hundred dollars. All of it is worth it to an attacker to steal. Organizations of every size have valuable information that cybercriminals want.

Cybersecurity is a silent killer. Most people are completely unaware that it's happening until it is too late. It's similar to how people can feel fine but have illnesses in their body. The problem is that by the time there are visible symptoms, the illness is terminal. By the time the organization or person has a visible sign of a cyberattack, it is frequently too late.

Often adversaries do not use your stolen information until a significant period of time has passed. For example, several years ago, there was the large-scale Equifax breach. Hundreds of millions of people's personal records—personally identifiable information and social security numbers—were stolen. Many people involved in the breach thought they were fine because they didn't see any problems right away. Yet we're still seeing, even in the last couple of months, information that was stolen four or five years ago slowly being utilized, and individuals and organizations slowly being targeted. In other cases, organizations have been compromised for several years and information is still being exfiltrated. Many people believe the goal of an attack is to steal information but, in many cases, the goal of an attack is to gain long-term access to an organization for monitoring and taking information at will.

When we talk about the concept of data theft, theft of information, and identity theft, we're not really using accurate terminology. The concept of theft means someone took something from you. If I stole your car, you'd no longer have your car. You would know your car was stolen. But it's not really identity *theft* because they're not taking it from you—they're copying it. With cybersecurity, someone can steal your identity, bank information, and passwords, and you'll never know, or if you do, you'll find out when it's too late, because it's technically still there and they just took a copy of it. Your information can be sold on the black market without you knowing, and it can put you in debt, hijack your identity, and slowly destroy your life. The problem with this type of cybertheft is that most people are not immediately aware that something happened. If someone stole your car, you'd realize immediately that it has been taken; however, when your identity is "stolen," you could be completely unaware for years. It's not until you've suffered a major loss that you realize what happened. With many of the large-scale tax scams that I have worked on, the social security number was stolen years before, but the adversary waited for the right opportunity to exploit the individual. For example, in one case, a person owned their company, but when his identity was stolen, he realized he was not making any money. It was not until three years later, when he made a lot of money and was going to receive an $800,000 tax return, that the adversary issued a false record and stole the money. Today's adversaries are patient and very clever when they launch their attacks.

Well, you are probably thinking that this can't happen to *you*. Your logic is that this is something that happens to other people. Because why would a hacker target you when they could target a large organization like Marriott, right?

Wrong. The problem is, it's not a matter of *if* it's going to happen to you . . . it's *when*. The number one problem organizations face in dealing with cybersecurity is they do not think they will be a target. One of the major themes of this book is for you to recognize that you, your company, your family, and your country are targets and will continue to be targets.

MY LIFE AS A PROFESSIONAL HACKER

Let me begin by telling you about myself. You may be wondering how I got here. I took an interesting path toward becoming one of the foremost experts in cybersecurity. In the 1980s, I was about to graduate high school when one evening, a good friend of the family came over for dinner. He asked what I wanted to major in at college, and I expressed my interest in architecture, because I was fascinated with buildings, bridges, and other structures. He said to me, "Eric, everything is going to computers. Why don't you major in computer science? And if you know about computers, how they work, how they operate, and how to program them, then you can get a job in any field that you want."

That seemed like good advice to me, so I went to college for my bachelor's degree in computer science. After a couple of classes, truthfully, I wasn't sure if I liked it. At that point in time, computer science was really electrical engineering with a little programming added into it. I thought maybe I was missing something, so I wanted to get a job to see what a professional computer scientist actually did. I decided to go to the school's co-op office to look for an internship.

When I went in, they smiled and said, "Your timing could not be more perfect. We actually have the CIA recruiting tomorrow on campus, and we have one slot left—it's yours if you want it. We've been calling around trying to fill it, and had you come an hour later, it probably wouldn't have been available."

I said, "Absolutely."

I showed up the next day and the office gave me an overview, but they said, "Eric, this is a very high bar, and we haven't had time to really prep you on interviewing, so look at this as an experience, but don't expect a whole lot out of it."

I went into the interview, and twenty minutes later, they gave me a big folder. When I walked out of the interview, everyone in the office was looking at me funny, and one woman finally said, "Oh my God, what did you do?"

I blinked. I said, "I don't know, what did I do? Did I do something wrong?"

She laughed and said, "No, no, that's good. That means you've got a full application and you did really well. We just did not expect that."

I had to fill out that huge application going back four generations in my family, and this was before the internet or Google. It was a great exercise to sit with my parents, get all the details, and fill it all in. And then I mailed it back and waited.

Four or five months passed, the semester ended, and I figured I wasn't going to hear from the CIA. I took an internship at Grumman Aerospace (now Northrop Grumman), and I worked on the security of the radar system for the F-14 fighter jet. I even got some experience playing in the simulators and working with the planes, finding out how they operate.

Three-quarters of the way through that internship, I got a call from the CIA, offering to fly me down to Northern Virginia to go through all the background checks, the polygraphs, and the rest of their testing because, as an intern, you need the full top-secret clearances.

I jumped. I flew down for my first time to Washington, DC, did everything, and again, waited.

A few months later, they called to tell me I was clear, and they wanted to fly me down to assign me to an office. What I didn't realize is that, as a co-op (part of their cooperative education program), you have full clearances but you're a free resource to the office that hires you. Co-ops are paid from a different office, and you don't go against head count. An office might only be able to hire four people, but they could have as many interns as they wanted, which meant offices at the CIA loved interns. It was actually a reverse interview—they were all trying to convince me to work for them.

I went to different offices—a networking operations center, operating systems, programming—and one of them was the office of security. I talked to my advisors, and most of them pretty much said, "Eric, go with networking. That's where the future is. Everything's becoming interconnected. Networking is what you need to know. Security is a fad—they're just going to figure it out. It's not really going to be a big, growing field. Just focus on the networking side." I clearly don't follow directions very well because I decided to go with the office of security.

During my time at the agency, I would test the security of different systems, including pre-release versions of operating systems, to see if they were properly locked down and protected. In one case, I was able to recover deleted information off a hard drive and specialized microscopes. And I remember one of the moments that sort of changed my life. At CIA headquarters, there's a big auditorium called "the bubble," where they have huge meetings and presentations. There was an all-hands meeting for the directorate that the office of security was a part of, and they were talking about the new (at the time) internet-based systems that I was involved with, which we were going to start rolling out. At the end, when they asked if there were any questions, I raised my hand (despite my boss mouthing at me from the front to put my hand down).

I asked, "How do we know these systems are secure if we're putting them out on the internet where anyone could get access to them?"

Everyone looked at me and then at each other. Finally, one of the directors said, "Eric, that's a really good question. Why don't you solve that for us?"

I figured that there would be some sort of mathematical operations you could perform to determine the overall security. It turns out, you can't. You can't prove a system is secure. You can only prove that it's not secure by trying to break into it. That was when I launched a seven-year journey during which I essentially became a professional hacker for the CIA, learning how to break in or compromise just about any computer system that's out there. I worked at the CIA for ten years total.

I later helped to build The Sytex Group, Inc. (TSGI), and when it sold to Lockheed Martin for $400 million, the president and CEO kept me on as their chief scientist to directly report to him. I was on call 24/7 to respond to incidents, like when their F-22 got hacked by China and I had to fly to Texas at 3:00 AM to handle it.

McAfee brought me on as their chief technology officer to redesign their portfolio to make it more current and align to customer needs. Two years after I did that, we sold to Intel for $2.3 billion. I've worked for eleven years for various entities, including the Gates family and several other high-worth people. I was also a commissioner for cybersecurity for the forty-fourth president—President Obama. I've written tons of technical books for specialists in my field, and I was inducted into the Infosecurity Hall of Fame, which picks one worldwide expert a year. When companies worldwide need cybersecurity, I'm at the top of the list. When an international company was looking to go public and had a major breach five years ago, and Wall Street had concerns about the security of the company, they hired me to work with the CEO and executive team to train their security team and raise overall awareness.

I've appeared on major news outlets like CNN and *60 Minutes*, and in publications including *The New Yorker*, as the expert every time there's an attack that the media reports. No one in cybersecurity has more experience and has the insights into effective security solutions as I do. Now, as the number one cybersecurity expert in the country with over thirty years of experience, I want to teach you about the realities of cybertheft and hacking and what you can do for yourself, your company, and your family. I'm going to give you an inside look into the current cybersecurity war, the places you are vulnerable, the solutions we can take to help stop cybercriminals, and what to do when you experience an attack. Let's go!

ONE

WE ARE ALL TARGETS

Whether we like it or not, or whether we want to admit it, we spend much of our life living in a virtual world. Even before COVID, many people shopped online, interacted online, and attended classes online. With the epidemic, most activity has moved online, with minimal or greatly reduced social interaction. People are doing virtual happy hours, and many schools are conducting a majority of their learning from kindergarten through college online—and this trend is going to continue. The problem with living our lives online in cyberspace is that people believe it is safe. The top comments I hear are: "Why would anyone want to target me?" and "The software and applications we use are secure." These are the two biggest fallacies. If you want to survive in cyberspace, there are two core rules you need to remember: 1) You are a target and 2) Cybersecurity is your responsibility.

Just think about how connected we all are now. At work, you're very likely on some computer: Even if your actual job doesn't demand it, your employer has your social security number; they may have your medical insurance details. In your personal life, you may post on social media. You likely have a smartphone; you may have a tablet. You may have a smart TV, a home security system tied into the web—even your kitchen appliances may "talk" to the web. You are at risk everywhere you turn. And where does your stolen information end up? The dark web.

THE DARK WEB: IT'S REAL AND THEY ARE AFTER YOU

Your information is an incredibly valuable commodity on the dark web. When you hear "dark web," you're probably thinking it's something out of Tom Clancy's Jack Ryan novels, right? But it is real—and dangerous.

The dark web consists of servers on the internet that are controlled and managed by criminal elements. You actually need special software in order to access them—called "onion routers"—and you need to have special passwords and special access. In some cases, you have to commit crimes in order to get access to a sensitive area. While breaches and hacks of personal information occur on the internet, the dark web is where it goes to be sold.

If someone has your full identity—your full name, address, social security number, phone number—that sells for approximately twenty-two dollars on the dark web. Think about how many identities are in a data breach and start doing the math. That's serious money. For example, the Target breach was 41 million records, Equifax was 145 million, and Marriott was half a billion! While there have been and will continue to be major breaches, the important thing to remember is that if it is your information that is stolen, or your bank account that is wiped out, whether it was one record (yours) or one hundred million, if it impacts you, the breach is real.

While the media focuses on the number of records, this is only the beginning. Ultimately, the cybercriminal wants to do something with the information and/or make money. Therefore, the probability is that the stolen information is going to be used in a criminal manner, like opening up fraudulent credit cards or credit lines or targeting your bank information. I know companies that will pay for information on the dark web to add to their mailing list or get the emails and phone numbers of people within a certain demographic.

As mentioned in the Introduction, your stolen identity or bank information, bought and sold on the dark web, may not necessarily be utilized right away, so you don't have any visible way of knowing that it has been compromised. Also, while there are companies that monitor and watch the dark web for such illegal activity, in many cases they do not have access to the very sensitive areas because criminal background is needed for that access. Think about how people join gangs: You have to kill somebody to prove that you're not a cop. Many of the areas of the dark web buy and sell data and information in the same way. You either have to commit a crime or show that you have 50,000 real identities in order to get access to the information in that data. Now, of course, there are organizations like mine with indirect connections, so we can still monitor and see what's going on, although it's nearly impossible to monitor everything.

However, a particularly important point to bring up is that many organizations that suffer major breaches do not have accurate logs, so they do not actually know whose records have been stolen. For example, they know that one hundred million out of eight hundred million records were compromised, but they are not able to determine which ones, because they don't have proper security measures in place. This is why, after a breach, you might receive a letter that says: "There is a possibility that your information might have been compromised . . . " I know it frustrates people that the company is being vague and won't actually tell them the truth, but the reality is that they don't know.

There may be certain rare cases when you want to hire someone just to monitor the information for a short time. I worked with a major corporation that was about to go public, and when that happens, there's a four-day window when there's a lot of sensitivity. During those four days, I hired a team to monitor the dark web to actively look for any sensitive information and take proactive action to minimize any damage. If something came up, I would put delays in place. If you have a small window, dark-web monitoring is helpful, but otherwise, assume cybertheft is happening and put precautions in place to combat it.

You should begin with the assumption that your data is out there and act as if you've already discovered that. I asked a business that question: What would you do if you found out your company was compromised? This executive said they would rescan the servers, make sure there was no exposure, and put on more staff to monitor activity. I told him that he should do that now. It will cost less before any theft occurs than if there is a bounty on the information to get it back.

The dark web is a real threat, and you and your business are targeted. The sooner you wake up to that, the better you'll be able to prepare. Act as if your data is compromised and be proactive. It is better to assume your data has been compromised, take aggressive action, and be wrong than to take no action at all!

I see two big problems when I talk to executives about cybersecurity. The first is that nobody thinks they are a target. Whether they're small start-up companies, entrepreneurs, even large oil and gas companies, one of the most prevalent comments I get is that they think no one's going to target them. For some reason, people believe that cyberattackers only go after the government or governmental entities. The fact is that many of the nation-state attacks, from North Korea, China, Russia, and Iran, actually target small-to-medium-sized businesses and individual businesses. The reason is simple: They're an easier target.

I was chief scientist at Lockheed Martin, an American global aerospace, defense, security, and advanced technologies company, for many years. And yes, they were targeted. You might have even read that they had some attacks from the

Chinese military with the Joint Strike Fighter, because they got hit on a regular basis. So let's think about this: How hard do you think it is to break into Lockheed Martin? Now remember, this is a company that knows they are a target. They recognize that they are a large government contractor, so they spend hundreds of millions of dollars on cybersecurity, and they have hundreds of people monitoring and watching their network, 24/7.

Foreign adversaries on a regular basis prove it can be done, but it's fairly difficult to accomplish. Now, what if you have a small company that doesn't consider itself a target? They might spend $200 on endpoint protection and maybe have a half a person doing some basic security.

Which company is easier to break into? Of course, the smaller one.

An adversary knows this. They could go after a super large company or bank and break into their database, which would take a lot of work and energy to steal thirty million identities with a high chance of getting caught . . . or, they could target thirty million individuals who don't think they're a target and steal their information without hardly anyone noticing. Which one do you think is easier and quicker? Of course, the latter, going after the individuals.

This is one of those counterintuitive points that many successful people and businesses overlook. They have valuable information, and because they don't believe they're a target, they're not investing a lot in cybersecurity. It makes it so easy to go after them. And that is exactly what the adversary is doing—to you.

It's why I've written this book—to raise awareness about the major prevalence of cybersecurity threats and the risks you face. The problem is going to continue to get a lot worse before it gets better because more and more attacks are going after individuals, since it's easier and more effective.

FUNCTIONALITY VERSUS SECURITY

Another misjudgment or assumption that people and companies make is believing that cybersecurity is somebody else's responsibility. I bet you think that if you buy a cell phone, it'll come secure. If you're buying a Microsoft operating system, Microsoft should secure it. If you're using third-party cloud services like Salesforce and other software available via the cloud, you shouldn't have to worry about it. Right?

I sit in so many executive awareness meetings with major, powerful executives who look at me and say, "Eric, I am running a $200 million business. My company has an IT and security department. That's their problem. They should have to deal with that, not me. Why am I dealing with this issue when I have a huge business to run?"

Yes, Microsoft, cloud providers, your IT, your security department—they do give you a pretty good, safe computer. But ultimately, the safety is based on the operator.

Let's look at a quick example: You can buy the safest car on the market. It has all the airbags, antilock brakes, the latest in all the safety features available. Can't you still take that car and drive it at a hundred miles an hour into a brick wall? Can't you injure or kill yourself in that very safe car? Absolutely. The safest car on the planet doesn't matter if you don't have a safe operator. And that's the missing link in cybersecurity.

Your company IT departments, your cloud providers, Microsoft, they are all giving you a safe car. But if you use it recklessly or even operate it correctly, you can still get hurt, injured, and be a victim.

If you put a car up on cinder blocks in your garage, I think we can all agree that's a very safe car. If you never opened up your garage, and you never took it out of your garage, and you just sat in that car every day, the probability of getting into an accident is almost zero.

But what's the point? Keeping a car in the garage is not valuable to you. In order for your car to provide value, you have to get out and drive it. As soon as you add in that functionality, the security or safety starts dropping and gets lower and lower. It's the same with technology—100 percent security doesn't exist (if you want to use the technology). No matter what you do, no matter what you put in place, you can't be 100 percent secure . . . unless you want to keep that computer in a box and never use it. As soon as you start using a device or add in any functionality, security drops below 100 percent.

I used to have a saying that the only way to be 100 percent secure is to go Amish. Driving a horse and buggy and working by candlelight makes it pretty hard for somebody to hack you. The irony is that it's not true because the Amish still have to pay taxes, and the Pennsylvania tax authority actually got hacked and records were stolen, including those of Amish taxpayers. So even if you live an Amish lifestyle and you don't use computers, but you're sending paper records to entities that put your information in an electronic format, you could still potentially be a victim.

I was giving a presentation where a slide was shown that said "100 percent security doesn't exist." I've had that slide up in hundreds of presentations, but for some reason on this particular stage, I don't know why I did it, I stopped and I looked at the slide, and said, "Is that really true?"

People in the audience started debating and arguing with their neighbors. So I held up my cell phone and I said, "Okay, let me ask you a question. Can I make this cell phone 100 percent secure?"

Once again, people debated, and somebody finally yelled, "Well, turn it off."

I said, "Great."

Then someone yelled, "But if you turn it off you can turn it back on again. So smash it with a hammer."

I said, "Okay, if I take the phone, turn it off, smash it to pieces with the hammer, can we all agree that it's 100 percent secure?"

A guy in the second row said, "You could glue it back together, so you need to light it on fire."

I said, "Okay, if I take a phone, I turn it off, I smash it to pieces with a hammer, I pour lighter fluid on it, and I light it on fire and burn it to a crisp, can we all now agree that that is 100 percent secure?"

Everyone agreed and seemed satisfied.

I said, "Then why even have a phone? The problem is 100 percent security means it has zero functionality and zero value."

Imagine we have a sliding bar with 100 percent security on one end and 0 percent functionality on the other. And then as you slide the bar down, at the very bottom, you have 100 percent functionality and 0 percent security. Every time you add functionality, you're decreasing the security. They have an inverse relationship. So the more functionality you add to your network, application, etc., the less security you have overall.

While you can't have complete security, there is a lot you can do to protect yourself that will make you less of a target. We'll cover that in detail throughout this book.

The first lesson is in functionality. Most executives ask themselves before they make a decision: "What is the functionality, and what is the benefit?" If the functionality or benefit is good, they'll do it.

For example, even though your company may say they don't allow USB drives to be plugged into their computers, leaders can and do override that. When an employee comes to you and says, "Listen, I can increase my productivity by 30 percent if I can go in and have USB access," you decide that a 30 percent increase in productivity is pretty great, so you sign off on it. But here's the problem: You didn't ask the second question—and nobody does—which is: "What are the security risks or security exposures by doing that?" That's the lesson I really want you to learn.

Let me give you two scenarios. In scenario one, you can increase somebody's productivity by 30 percent, but that person has been known to violate policies, has failed different tests by the company to see how well an employee follows the security policy, and isn't very security aware. So you can increase productivity by 30 percent, but the overall risk of that employee losing the USB and the data being stolen is 90 percent. Is that a good decision? Probably not.

On the other hand, if this is a really safe person, they're very aware, they don't have any security violations or issues or any other red flags, and you can increase their productivity by 30 percent while only decreasing your company's security by 5 percent, that's probably a good decision.

But you only know that makes sense by asking that second question. When you ask the first question—"What is the value and benefit?"—be sure you also ask the second: "What is the security risk or exposure?"

I love reading, and recently I was reading some books by Warren Buffett, the incredibly successful investor. His strategy is to invest by looking at the downside and then reducing it. He says the reason why so many people are bad investors and lose so much money is because they only focus on the upside. That's why Warren Buffett doesn't like Bitcoin. He doesn't like cyber currency because yes, the upside is huge, but the downside is disastrous. You could lose everything, and many people have lost everything with cyber currency, hence he doesn't recommend it.

That's the exact same strategy I've used in cybersecurity for over fifteen years. You need to think like Warren Buffett.

One of the things I get asked about the most these days are home devices, like Alexa, toys that have internet connectivity, or IP video cameras. People want to know if they are secure and if they should get them.

It's not a yes or no answer. It's not even the right question. You have to ask yourself: What is the value and benefit versus the security risk? If the value and benefit exceed the security risk, then you should do it.

I'm not always at my office or have employees working there every day. There have been occasions when packages were delivered and left outside for ten to twelve days, and it rained and snowed, and by the time I came home, the package was destroyed. It's also a very busy residential area in a commercial development, so some packages that were delivered disappeared because somebody took them. I now use Amazon home delivery, where their authorized people can open up my door, put the package inside, and lock the door again.

You're probably shocked that a security professional would even consider such a thing. But it's simple: The benefit I get is that my packages are safe and secure. They aren't destroyed or stolen, and I know that they're waiting for me when I get home. Yes, Amazon does have an access code, but every time they enter to deliver a package, I get a notification. Amazon home delivery requires that you have a video camera inside, so when the courier opens the door, I can watch them come inside. And the Amazon home delivery people are allowed to place the package inside your doorway, but they are not allowed to step into your house.

Could an Amazon person decide to come into my house and rob me? Yes. But I would have it on video. Both Amazon and I would know about it because the delivery person came inside and didn't lock the door quickly. Somebody could also rob my house anyway. So to me, the benefit outweighs the security risk, and therefore I decided that it's a good service to use.

The problem in cybersecurity is very few people ask that second question. So they'll put an Alexa in the house, but never look at the security risk or exposure because they believe that's someone else's problem to consider.

It's that thinking that puts people in harm's way. We believe that services and devices are innately supposed to keep us safe, so we don't consider if that might be true. But it is true and it's something you need to consider, especially every time you're adding functionality to technology.

A GLOBAL PROBLEM

While the challenge of whether to put an Alexa in a house is a very local problem, the challenge with cybersecurity is a global problem. Countries and criminals around the world are currently and consistently engaged in cyberattacks, and because there are no international borders, and it can be done from virtually anywhere, it is a low-risk, high-payoff crime.

The Russians are in our power grid right now. They have access to our technology, communication, and beyond. If you don't believe me, you can see for yourself. Publicly traded companies have to disclose in their yearly findings to the Securities and Exchange Commission anything that can impact the success of the company, and that includes cyberattacks.

If you look at Intel, which is one of the major chip manufacturers in the world, based in the United States, you can see they've had several attacks that significantly impacted their stock price. They put that right there in their stock holdings. A southern utility company, which has nuclear power plants, has experienced several cyberattacks that could significantly impact the value of their company, which they've put in their SEC public filings. And of course, we know about some of the more publicized ones, like Marriott and Saudi Aramco, who have been hit.

Even if you think this doesn't directly affect you, it's happening all the time. Many people do not read these public filings and, as I said, the media isn't really super keen on cyberattacks, so a lot of it just gets swept under the rug. But know this: The Russians do have access to several utility companies and nuclear power plants in the United States.

You're probably wondering why this isn't spreading panic. The answer isn't simple, but it's analogous to the Cold War. Russia had long-range inter-ballistic nuclear weapons that could have wiped out the United States multiple times over. On the other hand, the United States had hundreds of nuclear weapons that could take out Russia many times over. Both sides had some very scary technology, but nobody ever used it because both countries were smart enough to know that it would mean mutual destruction. That's where we're at now. Yes, the Russians are in our critical infrastructure and could cause major problems, but we're also in their infrastructure and can very quickly and easily return the favor.

Still, we are in an exposed state. I have worked for several presidents as an advisor on cybersecurity, and I find it interesting because every one of them has asked the same question when they take office: "Can you tell us all the internet connectivity points coming in and out of the United States? So if we wanted to, could we go in and monitor, block, or temporarily disconnect the United States from, say, Russia or China from an internet standpoint?" From what I can tell, no one has been able to tell them yes.

Here's the short reason why: The United States technically started the internet in the late 1960s/early 1970s by creating something called ARPANET. During the Cuban missile crisis, the military wanted to build a resilient command and control network so that if part of the country was destroyed, the network could still operate, and we could still be able to communicate back and forth with what was remaining of the United States. This eventually turned into the backbone and critical components of what we now call the internet. As a result, we are more vulnerable than other countries. North Korea, for instance, has approximately six connectivity points, so if they want to monitor, control, block, or launch a cyber war, they just need to disconnect six points to take their systems off the internet and limit all their country's private information. Russia recently disconnected from the internet for twenty-four hours, and I'm not going to lie to you, it scares me. It shows they're preparing for something. They wanted to see if they could protect and secure against cyberattacks. Russia claimed they did it to show that the country can operate independently without connectivity to the rest of the world, but we all know in politics that what is said in public is not always aligned with the truth.

Unfortunately, not only can the United States not do that, but we don't even know all our connectivity points. Here's the bigger problem: We don't know what we don't know. We don't really know how bad the problem is. We know about some attacks, but we also know, because of the visibility and monitoring, that we are not catching all of the attacks.

It's the attacks we're not aware of that really worry me. It's not just Russia in our infrastructure or North Korea that should concern us. I don't think it surprises anyone when I say that China is actively targeting and going after our infrastructure. But think about any technology, whether it's a computer, laptop, cell phone, printer—virtually any electronic device on the planet—and flip it over. What are the three magical words that you will see there? "Made in China." So if we know that the Chinese are actively going after us, what makes us think that they're not embedding malware at manufacturing time? Surprise—they are. We'll look at that more closely a little bit later on.

A PERSONAL PROBLEM

Still don't believe me that any of this could happen to you? Let me tell you, I am constantly dealing with personal cybersecurity attacks down the funnel on the personal level.

I knew a couple who had been married for fourteen years when they came upon a fairly significant inheritance—$1.3 million. This far surpassed any money they had in their bank accounts, any money they would ever make in their lifetime. They decided to buy their dream house with this money.

They found the house, fell in love with it, did all the work, made arrangements with the brokerage company, and they were finally getting ready to close. A couple of days before everything was completed, they received an email that said the sellers changed the bank where they wanted the buyers to wire the money. It included the new banking information along with a reminder that it needed to be sent forty-eight hours prior to closing.

They had been emailing back and forth with this person at the closing company for a while, so they knew who the person was and even wrote back to confirm the switch. The closer said, yes, this happens very often and is fairly common. They went to the bank and the bank even sent an email to verify.

They transferred the money. Two days later, they showed up at the closing, so excited to move into their new, beautiful dream house.

The closing agent told them, "We're going to have to postpone closing. You never transferred the money."

The couple was shocked. They told the agent they did. As the day progressed, their stomachs started sinking deeper and deeper as they realized what had happened. It turned out that a hacker found out about the closing, broke into the closing company's server, took over the mail account, and falsified emails to this couple. Because the couple only verified via email, of course when they forwarded or replied to the email—since it appeared to be the legitimate email account of the closing company—the attacker who gained control of the entire system was able to reply back.

The significant point here is the window of forty-eight hours. If you catch financial fraud or financial transfers within twenty-four hours, you can usually reverse them because they're in a pending state. But after forty-eight hours, the money is typically gone. The other problem is that, because the couple instructed the bank and the bank did exactly what they said, the bank is not usually liable. If your bank account gets broken into and somebody steals money that you didn't authorize or know about, then the bank is liable. But in a situation like this where you're telling the bank to do it . . .

This is not only a real story, but I can tell you thirty just like it. I get involved in these types of situations all the time.

I even had another case with Bitcoin, which is a cyber currency, in which somebody lost $7.2 million because a hacker went in, found out their cell phone number, was able to spoof the number and get their two-factor authentication code to be able to do the transfer. (Most currency transfers like this need that double authentication.)

In yet another case, an entrepreneur traveled to China to negotiate a $400 million buyout of a US company. Both sides negotiated all day and got nowhere. The entrepreneur went back to his hotel room and emailed and texted his American business partners on his phone and his computer: He told them that he didn't think the deal would go through if they wanted to stick with the price of $400 million.

His partners finally told him: "Listen, we will go as low as $280 million if we need to, because we really want to sell this. We desperately need the money, as you know, but try not to go below $340 million."

They emailed back and forth—with multiple parties representing the company to be sold—with this information before the entrepreneur finally went to sleep.

The next morning, he returned to the negotiations and started off by saying, "Okay, I know we talked $400 million, but we really want this deal to go through. So we're going to lower it to $360 million."

The Chinese counterparts started laughing and replied, "Listen, you know and we know that you'll go as low as $280 million. So unless you give us that deal, we're not taking it."

His jaw hit the table.

HOW CYBERATTACKS WORK

Let's look at how a cyberattack actually works.

First, a cyberattacker finds a target. To find the target, you have to exploit a weakness and take that information. If you are staying at a hotel and filled the hotel room with gold and someone wanted to steal your gold, there would be three pieces, and only three pieces, of information needed to do that: the address of the hotel, the room number, and a vulnerability or weakness to be able to get into the room.

Hacking really comes down to identifying a target, knowing where the data is, and exploiting a weakness in that target. As we work through this book, you'll see there are two general targets: a server and a user. So if a hacker wants to steal information, they can find a server that's out on the internet and discover a weakness in that server to exploit and break in and steal the data. That's what we've seen with many of these breaches.

A hacker could also target an individual. That's typically called a "phishing" attack, where a hacker sends you a legitimate-looking email, which you believe came from a valid source, so you're tricked or manipulated to click on the link or open the attachment. Once you do, they can infect your system or hold your data ransom.

THE THREE TENETS OF CYBERSECURITY: CONFIDENTIALITY, INTEGRITY, AND AVAILABILITY (CIA)

Cyberattacks are a true threat, but the fact that they don't often make it into the news, or are on and off the news so quickly, is why I'm really writing this book: to raise your awareness that this is a serious problem that we have to start addressing now. The main thing I want you to walk away knowing is that you are a target, and cybersecurity is *your* responsibility.

When I talk about cybersecurity and cyberattacks, it really comes down to what I call the CIA. I'm not referring to the place I worked: the Central Intelligence Agency. I'm referring to the three foundational tenets of cybersecurity:

Confidentiality
Integrity
Availability

These are the cornerstones of cybersecurity. Let's break down each one.

Confidentiality is to prevent, detect, and deter the unauthorized disclosure of information, which means making sure that information is not inadvertently disclosed. That secrets stay secrets.

Integrity is to prevent, detect, and deter the unauthorized alteration of information. This makes sure that information stays accurate and correct.

Availability is to prevent, detect, and deter the unauthorized denial of access to information. This means making sure that information is accessible when and where we need it.

When many entities think of security, they only think about confidentiality, that their information is kept protected. Nobody wants their identity or critical information to be stolen, or their private information or photos stored on their computer to be publicly available. That's all confidentiality.

However, we need to step back and realize the other two are just as important. If you're going in for an operation at the hospital, sure, you would prefer that people don't know about it, wanting confidentiality. But you do want to make sure the doctor performs the right operation and that those records are accurate. What if somebody can go in and alter the records and switch patients so the doctor performs the wrong operation? It sounds ridiculous, but this has happened and, unfortunately, more times than the medical community would like you to know about. So I would say in some cases integrity is probably more important. Another example of the importance of integrity is your bank account. While many people think confidentiality is what's important, in reality someone finding out how much money you have in your bank account is not really a major issue, even though you might prefer to keep that information to yourself. But if someone can alter that figure, that's a big problem.

Even more people overlook availability. You believe that cloud backups will make sure your information is always available. However, availability has become a major issue with what we call "ransomware attacks." That's when you get targeted, and you click on a link or open an attachment that you shouldn't, and all of your data gets encrypted, and unless you pay the ransom to the thief, you'll never get it back. You may think that would be okay because you have that data backed up. What you may not realize is if you get hit with ransomware, and your primary hard drive or data store gets encrypted, it will replicate to all of the other data stores, including the cloud, and all of your data will be held ransom until you pay that amount. If you don't, you'll never get your data back.

This happened with WannaCry, which included a ransomware attack on hospitals in the UK, an attack in which Britain's National Health Service was paralyzed. In this case, there were three main hospitals, and data was replicated. One doctor clicked on one link in one email, and 80 percent of all the patient records across all three hospitals were encrypted and held ransom. Those hospitals were actually out of business for four days, and patients had to be moved to other hospitals. The attack replicated that corrupt data, which made it a major issue. (So you can avoid that trap, later on I'll show you the difference between transparent and nontransparent backups.) But even if you think your data is backed up, you might not have what you think you have.

Ransomware has been around for twenty years, but it's changed. If you got hit with a ransomware attack two decades ago, attackers would encrypt all your data and charge a massive amount to give it back—let's say $40 million. How many companies do you think would actually pay $40 million? Very few. But the attackers didn't care because if they hit a bunch but only got one or two companies to pay $40 million, that was a pretty good return. The second problem with these older kinds of ransomware attacks is that they were mostly scams, and businesses wouldn't get their data back. So one or two companies would pay the $40 million, not get their data, and they would quickly tell other businesses not to pay the ransom. Those kinds of attacks died off very quickly.

Then a few years ago, ransomware attacks came back with two twists, one of which was a much lower price. It is important to note that while the above example focused on businesses, ransomware can also target individuals working on their home computers. Now, we see individuals and businesses getting their data stolen with the ability to retrieve it for the bargain price of $79 (individuals) or $50,000 (companies). If you're working at your computer tonight and you make a mistake, and you click on a link and all your data gets encrypted—all your pictures, tax records, documents, and information—and a message pops up that says you can get your data back for $40 million, you'd be pretty upset. But you'd let it go because it's an impossible amount you wouldn't even consider. However, if you're sitting in front of that same computer and you get a message that says you can get your data back for $79, how many people are going to do it? A large number. So the new ransomware says we don't need $40 million from one person—we need $79 from a million people. And that's why the WannaCry attack made over $15 million within a three-week period.

The second change that happened with ransomware is that for it to work, for the thieves to get a million people to pay $79, those users certainly had to get their data back. Otherwise, everyone would make a big deal and nobody would pay the ransom. Believe it or not, these ransomware companies now focus on customer satisfaction and customer service, and—I kid you not—they have tech support lines. They have a money-back guarantee. If you aren't able to recover your data or get your information back, they will give you your money back.

It's crazy, right? I would be at a party and someone would say to me, "Eric, I got hit with ransomware last week, and I paid the $79, and I tell you they were the nicest, kindest people. I got my data." These are thieves! But they were so nice and kind that people overlooked the malicious nature of what masquerades as a business.

These types of emails, where someone tries to trick or manipulate someone else, have been around for a while. Probably one of the first, most popular, and visible ones was called the ILOVEYOU virus, back in the mid 1990s. Recipients would get an email from a coworker that said "I love you" in the subject line. In the body of the email, it said, "This is a special message of love. Click here please." People would click on the email and it would infect their system and then would send the "I love you" message to everyone in their address book.

Now, in this case, it was purely disruptive. The author did get arrested, and his whole goal was to become famous, so he got exactly what he wanted. Today, attacks are directly focused on data, and if any cybercriminal gets caught, it's because they want to be. (Or they're incompetent.) If they know what they're doing, we'll never find them.

I remember working on ILOVEYOU. I was heading up security for a large telecom company at the time, and I remember thinking how it was obviously malicious. If you walk into work and there are eighty-five email messages from coworkers that say, "I love you," that is not normal activity—unless you work for a really screwed-up organization. People knew that was not legitimate but what did they do? They clicked on the message anyway.

While ILOVEYOU was clearly a scam, the concern is what happens when the email looks real. What happens when it doesn't say "I love you," but it comes from the IRS or your bank or FedEx, UPS, or your e-commerce site? That's where we are now. Today we are getting emails that look and seem legitimate, and one click is all it takes for all of your data to be compromised. As we get into our solutions later on in this book, I will give you some of the warning signs you can look for.

CHAPTER 1 REVIEW

Cybersecurity is a real concern, attacks are happening every day, and major breaches are occurring all the time that you're not hearing about in the news. It's going to happen to you, if it hasn't already. It's a global problem, business problem, and individual problem. One of the things you can do is think about functionality and security on a sliding scale. If you add more functionality, you enable less security. Cybersecurity is your responsibility, so before adding more functionality to your life or your business, ask: What is the functionality and benefit *and* what are the security risks? Remember that a cyberattacker only needs to identify you as a target, know where your data is, and exploit a vulnerability to get your information. Your vulnerabilities under the three tenets of "CIA" are:

Confidentiality, where somebody will steal your information;
Integrity, where they're going to alter or change your information;
Availability, where they're going to use ransomware or make your data unavailable to you unless you pay a certain amount of information.

These principles of cybersecurity are the kinds of risks you need to consider.

TWO

WE LIVE IN CYBERSPACE

We live most of our lives in cyberspace. When did it all happen? Many of you remember a time and place before cell phones, computers, and widespread technology. When I grew up, we still had a rotary phone with a cord in our hallway that we would use to make calls, and after school, I would run around outside and play with my friends. If I needed to talk to somebody, I had to pick up the phone and call them or write a letter.

I remember when I got my first job at Northrup Grumman. I worked for the radar division, which had over 120 people, and I was one of only four people in the entire department who had a computer. I needed one at my desk because of a project I was put on. In 1990, it was a top-of-the-line computer that cost $5,000 at the time, but based on today's standard, it was an old computer with a twenty-megabyte hard drive and no network connectivity. If somebody wanted to break into that computer, they would've had to get physical access to the computer system. Plus, it was located in a classified facility, making it even more of a challenge to break into. The ironic part is that the file cabinets that contained printed documents had far more sensitive information than any of the computers, so if someone broke in, the paper would have been the main target of opportunity. But I digress.

The changing landscape of technology also reminds me of when I worked at the CIA and was part of the virus investigation team. Back then and by then, everyone had computers, but they were not connected to the internet. There was no email or web surfing, and the computers that were networked were part of a closed, contained private network. If somebody got infected with a virus, there was only one possible reason, and that was because they inserted a floppy disk into their computer. If a computer's antivirus software went off, we asked that person a simple question: What floppy disks did you insert into your computer? We then traced back the floppy disk to see where it came from, identified any other systems that were compromised, and we were able to contain and control the problem relatively quickly. That is in direct contrast to where we are today with viruses. Now, malicious code can spread very quickly and come from anywhere, any place, any time, and be much more difficult to trace.

Even when people started using home computers with internet access, it was typically done via dial-up modems that usually utilized a third-party host such as AOL. In those cases, you technically were connected to the internet, but it was usually for short periods of time because you were tying up the phone line. If somebody wanted to attack your system, you were only connected for, say, forty-five minutes, which provided a short window of opportunity. Additionally, because the bandwidth was so narrow and the transfer rates so slow, there was only a minimal amount of information they could steal, and very little information was stored on computers in those days anyway. For instance, there was virtually no online banking, shopping, or social media, as we think of it today.

And then, bam! It seemed like overnight, everyone had multiple computers, cell phones, and other devices that were now turned on 24/7 with direct access to the internet. It's what makes it much easier and simpler for adversaries to break in, compared to twenty or thirty years ago. We're everywhere now, so they're everywhere now.

Think of the contrast today. Our cell phones are interconnected anywhere, any place, at any time. People can get access to us and our information whenever they want. We don't even realize it, but we have switched from living in a physical world to living in cyberspace. Think about it. Track your daily activities when it comes to work, your personal life, or even as a parent, and just look at how much communication is done via text, email, and social media versus how much is done face-to-face.

If your company got a contract, or something happened to a competitor, or a new product came out, or a customer was unhappy with your service, how do you find out about that? Most likely it's because somebody sent you an email, a text, or you read it about it on social media. We easily spend 80 percent of our lives in cyberspace. Communication, work, entertainment, research, you name it—it's online. Yet in many cases, we overlook the most obvious question: How is that information safe, protected, and secure?

CYBERSECURITY IS YOUR RESPONSIBILITY

Hopefully, Chapter 1 drove home to you that you are a target. No matter who you are, no matter where you work, no matter what type of company you own or you work for, you are a target. The number one mistake that I see when working with executives is they almost always view cybersecurity like they do accidents, illnesses, and robberies. *Those are things that happen to other people. Those are not going to happen to me. No one's going to target me.*

If you are online, if you have an identity, and you have a dollar in your bank account, which is just about everyone, repeat after me: "You are a target." As we discussed in the last chapter, today's adversaries are going after the lowest hanging fruit. And that is not the government or large corporations. It is *you.*

Now, how much money do *you* spend on your cybersecurity? For most people, it's probably fifty dollars a year for something like Endpoint Antivirus, right? How many people do you have on your cybersecurity detail? Probably zero. As mentioned, the media only covers the big stuff. When Jeff Bezos, currently the richest man in the world, has his cell phone hacked and compromised, sure, it's all over the news. But when other people like you have their cell phone compromised, which is happening all the time, that is not covered in the media. When I go into companies to perform assessments, we find hundreds of their executives' devices and cell phones compromised and they have no clue. That's never going to show up on CNN.

———

Your next big lesson is that cybersecurity is your responsibility.

"Oh, but Eric, our company uses a third party. We use the cloud. We're using Office 365. We've outsourced everything, so we don't have to worry about that."

Allow me to let you in on a little secret: You can outsource IT, and you can outsource functionality, but you cannot outsource liability. If you outsource to a third party, even if contractually they're obligated to provide proper security and they don't do it, you're the one who's liable.

If you remember, many years ago T-Mobile had a breach of hundreds of millions of client records, which we all heard about. But it turned out that it wasn't T-Mobile. It was actually their data center provider that had the breach and didn't implement proper security, but that part hardly made it into the news. People were angry with T-Mobile anyway because customers had given their data to T-Mobile, and it was T-Mobile they expected to protect and secure their information. Ultimately, you are responsible for protecting and securing information.

That's the unfortunate reality for you to recognize. You could be doing everything correctly to keep your business running, even predicting downturns in the economy, providing differentiating services, outpacing your competitors, but if you fail to protect your customer data, you fail to protect your critical intellectual property. Cybersecurity is one punch that can take out your entire business.

One of the things that has totally changed security from a business standpoint is the Target breach from 2013. It was one of the first really big breaches, so some people remember it, and a lot of people were directly impacted by it. What made the Target breach so significant is that prior to it, the playbook for companies when they got compromised was to simply fire the chief information security officer or whoever was responsible for security. Fire them, blame them, increase the security budget, hire somebody new, and move on.

But Target was one of the first breaches where the board of directors actually went in, held the CEO liable for not asking the proper questions and strategically worrying about security. The CEO was fired from the company a few months after the breach. That woke everyone up.

To effectively battle cybertheft and take the reins of your cybersecurity, you need to shift your mindset to accept that data theft is going to happen to you and your company. So you need to be prepared, and you need to accept responsibility. Our data is everywhere, and it's in the control of third parties.

I still have clients who will tell me they will only take us on to do security work for them if we will guarantee that they'll be 100 percent secure. I have to pass up that work because the only way to be 100 percent secure is to completely disconnect from the internet. Remember from the previous chapter, functionality

TARGETING TARGET

In 2013, during the holiday season, Target suffered a breach in which forty million credit and debit cards were stolen. It was estimated that the total cost of the breach has been over $200 million. This was one of the first massive breaches that started to get people's attention on the impact that a lack of cybersecurity could have on the business.

This breach was caused because someone at the HVAC company that Target used was compromised, and because the networks were interconnected, the cyberattacker was able to gain control of the HVAC vendor's network and use that to access all of the POS (point of sales) systems at all of Target's retail stores.

This is an example of not controlling and managing access to critical data and not following a principle of least privilege, in which only the minimal amount of access is given to third parties.

and security are on a sliding scale, and every time you add functionality, you're decreasing security.

If we look at some of the recent breaches that happened where hundreds of millions of customers' records were compromised, it's because the company decided that they wanted to have a system that contained critical data be accessible from the internet. Anyone could access that system, and it contained critical customer data that wasn't properly protected and secured.

That is an example of an unnecessary risk, where the risk outweighed any functionality benefits.

There are many organizations that allow people to access information from the internet in a way that's less risky. One primary method is a "multitiered architecture," where you have a front-end web server that's accessible from the internet, which doesn't contain any critical data, and there are multiple steps you have to go through in order to access the data. You're still allowing internet functionality, but you're doing it in a way that mitigates and reduces the overall risk to be able to verify, validate, and have early detection into what's occurring within that environment.

It is *always* a trade-off between adding functionality and decreasing security, whether it's in your business, with yourself, or at your home. Understand what the risk exposure is and ask yourself, "Is this worth it?" If you start asking those questions, you will enter a new security mindset. Any company that tells you they can prevent all attacks, that they can make you 100 percent secure, is lying to you. Don't believe them. A breach is going to happen, whether you like it or not. My goal when it comes to cybersecurity is not to prevent all attacks, but to control the damage and detect attacks in a timely manner to minimize the damage.

Before you get too upset and throw up your hands and say, "If you're telling me that no matter what we do, no matter what we put in place, we're going to be compromised and have a breach, why even bother?" keep reading.

PERSONAL SECURITY

We're taught about physical safety from a young age, like how we should look both ways before crossing the street or not to take candy from strangers. But no one talked to us about cybersafety when we were young, and we've only started talking to kids about it today.

You probably understand some of the basic components with safeguards such as passwords, but even then, how many people actually follow that? A survey recently came out where most people reported that they use the same password for all their accounts. So if somebody finds out the password for one of their accounts, they can get into their work, bank, e-commerce— everything. Survey takers also admitted their password has some relationship to them and typically contains a dictionary word as the basis or the only component of that password. We're in a very, very vulnerable state.

One of the things I can tell you, having been a professional hacker at the CIA, is if a computer is accessible from the internet, it's hackable. It's really that simple. So our cell phones and all these devices that have all this wondrous, amazing functionality also have security exposures and risks.

Go online to Google and type your name. If you have a common name such as Eric Cole, you might want to put some parameters in there. For example, if you search on Eric Cole, up until recently you would get three hits. You would get Eric Cole, a cybersecurity professional, an NHL hockey player, or a guy who's wanted in thirty-seven states for drug possession and selling illegal drugs.

Hopefully, you can figure out which Eric Cole I am. I know I'm definitely not the hockey player. The other two you might go back and forth the more you read this book, so I would put in "Eric Cole, cybersecurity." Or I might enter, "Eric Cole, Secure Anchor" (where Secure Anchor is the name of my company). I could even put, "Eric Cole, CIA," since that's where I spent several years working. But the bottom line is you would be shocked and surprised at the information that's out there about you.

Whenever I do executive awareness sessions, and I ask for a volunteer and put their name into Google, they're always amazed and dismayed at what I can find out. I can determine where they lived, information about their spouse, their kids, etc. I remember in one case, the person got very, very upset at me because on the screen we were displaying information about his children, their grades, and he started yelling at me that it was a violation of his personal privacy. I wasn't hacking. I was just Googling.

Even if you don't have social media, you still have a digital footprint. I often talk to people who will say, "Oh, I don't have social media. I don't have Facebook, Instagram, any of that stuff, so I am safe and protected." Actually, you're not. Because guess what? Third parties store information about you, your company stores information about you, and entities store information about you. The reason why it's very important for you to Google yourself is to understand your visible profile out there—because that's what attackers use to target you.

BUSINESS SECURITY

As part of our digital bodyguard services for our clients, we can predict with 93 percent accuracy who in the client's company is going to be targeted by cyberattacks, and who is going to either be phished or directly targeted for data theft, solely based on public profiles and information that's out there.

In one of the recent cases that we worked on, a CEO of a very large company—we'll call him Bob—was giving a keynote address at an event in San Diego at the Hyatt, and it was posted on the conference's website. It was public information that he would be speaking on February fifth at 9:00 AM—easy to find on the web.

At 9:05 in the morning, the attacker called the CEO's office, and the assistant picked up. The attacker said, "Hello, this is John at the front desk of the Hyatt. Bob checked in, but we had a problem running his credit card. It didn't actually go through. If we do not get a credit card within thirty minutes, I'm sorry, but we're going to have to cancel the room."

What would any good assistant do? Provide the credit card information because Bob is in the middle of a keynote speech, and she doesn't want her boss to lose his hotel room for the night. The attacker gained Bob's company's credit card information.

We've now trained that staff, and in that situation, we suggested that the assistant should ask for the extension of the person who phoned and call them right back at the hotel. If the person says, "Oh, absolutely, I'm extension 505. You can find our number on our website," it's probably legitimate. If the person says, "No, we can't do that. I have to do it now," or "I'll call you back in five minutes," and they start giving you the runaround, you can quickly see that most likely this is a scam.

The other interesting part, and one that's counterintuitive, is I often have a lot of high-profile executives say they want to stay off social media and then they'll be better protected. Actually, it's the opposite, and you're at more risk. The reason is that if you don't have a legitimate account, a hacker can set up a bogus account for you, and nobody would know. Just think: How easy is it to get a picture of you? How easy is it to find out basic information and details about you? The hacker can go in and set up an account as you. The hacker can build up friendships, contacts, and potentially use that to trick people into thinking it's you and get information from those connected to you.

Now you might say, "But Eric, social media doesn't authenticate, so somebody can do that anyway." Yes, but here's the difference: If there's a real Eric Cole account with my picture, and there's a fake one with my picture, then I can at least be aware of it so I can tell people and try to take action to get it removed. And if someone sees two accounts for me, they're going to look closely and ask, "Which one is legitimate and which one isn't?" I recommend having a social media presence, even if it's minimal, to better monitor and prevent bogus accounts.

We will continue to talk more about social media and those dangers, but I'm just trying to make you aware of how much of your world is in cyberspace and how much of your information is online. Just think for a second: What is the value of your laptop? The laptop that you work on every day, on which you might go in and check your bank account, maybe you file your taxes and do work, what is that laptop worth? A few hundred bucks? A thousand? No, that's the cost to buy a new laptop. But what is the cost of the data that's actually on that computer?

When I travel to certain countries like Russia and China, I know the probability that I'm going to be potentially targeted by the government based on what I did, what I currently do, and various other factors. I've learned it's much simpler for me, when I travel to those countries, to bring a throwaway laptop. We actually have a stack of laptops in our office that we buy for $300 each, and every time I go to China or Russia, I just take one of those, use it for two to three weeks, and toss it when I get back.

I know you're thinking that sounds extreme, but consider this: If I'm traveling in China, it's almost guaranteed that I will be targeted. So that means I would have to spend at least two to three hours of my time or somebody else's time going through that laptop. Even if we use a billing rate of $500 an hour, that's $1,000 to $1,500 to examine, and it's not a guarantee; yet, a new laptop is only three hundred bucks.

I urge you, if you go into those countries, using that type of throwaway laptop and a burner phone (which is a disposable or one-time use phone) could be incredibly valuable.

In fact, an executive from one of our client companies recently visited China, and it turned out his laptop was compromised, and now that client is using throwaway laptops as a protective mechanism.

A few years ago, before my son went off to college and when I traveled internationally, he would always take me to the airport. Once, when I was going to China, I left my regular laptop in my office and took a throwaway laptop. Before we left for the airport, my son saw my laptop still in my office, so he asked me whether I forgot to pack it. I said, "Oh no, buddy, because I'm going to a certain country, I'm not bringing it with me, and I'm bringing a throwaway laptop." He said, "Well why aren't you bringing it with you?" Without even thinking—I was just joking—I said, "Because it's a million-dollar laptop." On the way to the airport, as he was driving, he was quiet until he finally asked, "Dad, is that laptop really worth a million dollars?"

It got me thinking. On my laptop, I was working on three expert witness cases; each one of them had potential settlements of $30 to $40 million. I then had results from assessments and penetration tests (this is where you simulate an actual attacker and try to break into an organization) for several large banks. I kept going through the list, and I said, "You know something? I'm probably missing a zero. The amount of data and information that's on that laptop is probably $20 to $30 million, and I just never stepped back and really thought about what that value would be to somebody."

So I ask you: Do you have a million-dollar laptop? Do you have a ten-million-dollar laptop? Many of us are carrying sensitive data, especially if you're in the business world. And on the dark web, there are people who will buy laptops from corporate executives. They'll often pay anywhere from $400 to $500,000 for a laptop that's unencrypted, so they can access the data of executives of top businesses like banks, oil and gas companies, healthcare firms, etc. because they know how much that information is worth.

IT'S LESS LIKE GETTING ROBBED, MORE LIKE GETTING SICK

As I said before, not everyone in their lifetime will have their house robbed, but everyone will have their information compromised. It is similar to illness—everyone in their lifetime will get sick. This is why we need to stop thinking about our identity as something that might be stolen and think more about it as an inevitable health hazard.

No one on this planet is the perfect specimen of a human who is 100 percent healthy. No matter what you do, no matter what you eat, no matter how you exercise, at some point in your life, you are going to get sick. We all know that. We accept and recognize that. Your approach to preventing illness is to engage in activities like eating healthy and exercising. But if it's true that, no matter what, you'll get sick at some point, then why do we do all those things? Why don't we say "screw it" and just do whatever we want?

The reason we put all those measures in place is to reduce the frequency of getting sick and to minimize the impact that illness has on our life. If you get sick once every two years, and you miss one day of work, nobody's going to say you're not healthy. On the other hand, if you're getting sick every week, and you're missing two to three days of work, then yes, people are going to start saying that maybe you have health issues.

It's the same thing with cybersecurity. If you have a breach every one to two years, that's not an issue. If you can detect it, and there are only fifty records compromised, that's good. That means you're doing your job. On the other hand, if you get breached every single week and have millions of records compromised, you're probably not doing a good job with cybersecurity.

I have clients who tell me, "Oh Eric, we haven't been breached in two years," and I just cringe. If you're telling me that you haven't had a breach in two years, either, one, you have been breached but you haven't detected it, or, two, you are doing magical unicorn-grade security and are protecting yourself in a way that no one else on the planet has figured out. Trust me, it's option one. If you say you haven't had a compromise, it means you're not looking in the right spot.

I was recently in the UK giving a keynote address. Since my return flight wasn't until later in the day, when I was finished I decided to stay at the conference and listen to some of the other speakers. The next event was a group of panelists, and each panelist proceeded to introduce themselves. The first person, a chief information security officer, got up and said, "We know security. We haven't had any breaches, any compromises, any attacks in over two years." He was fist pumping, and everyone was cheering for him.

Then the next panelist got up, took a deep breath, looked around, and said, "We have approximately one to two breaches every year, and we usually catch them within three or four hours. It's usually just a few hundred records." You could hear the room. It was almost like people were going to start booing and throwing tomatoes at the speaker, and the panelist was almost ashamed of what he said. I was sitting in the back of the room, watching this dynamic play out.

I'll be honest with you, it was the most galactically stupid thing I have ever seen in my life. The person who claimed that they had no attacks, no breaches, no compromises for two years should be fired for gross negligence because that's not possible. If you're on the internet, if you're actively and aggressively doing e-commerce, marketing your products, that is impossible. It means they haven't been detecting breaches or looking in the right spot.

The person who said, "We've caught attacks. We've contained them within a few hours and controlled the damage"—that person and people like him are heroes. Those are the people that you want. I know it sounds counterintuitive, but if you're a businessperson, an individual, or an executive, and your security people are telling you, "Oh, we're fine. We're secure. We have no attacks, no breaches, no compromises," you should be extremely concerned and even frightened, because most likely it's happening and they're just not detecting it. I'm not just talking about the big businesses either. This is happening to everyone, at every level of business, and to individuals who are unaware of the violations and the fact they're going undetected. That is a huge problem in my industry. There are entities like large hotel chains that spend over $30 million on security every single year. They have over three hundred people on their security team, and those people are very well trained, and they were attacked for over three years before they detected it. Three. Years. And that example is a real, high-end company with top-notch security people.

The problem with security today is that many people are not looking in the right spots. Compromises are stealthy and blending in with legitimate traffic, so they are going undetected for a long period of time. Here's the scariest part: If you look at most of the major breaches that happened over the last several years, almost all of them were detected because of IT performance issues. The attackers broke into the database server and started stealing one hundred records. That blends in. Nobody catches it, so the attackers get greedy, and they start stealing one thousand records. No one catches them, and they get more greedy. They start stealing 10,000, 100,000, and because they get greedy, eventually it impacts the performance of the server.

When IT looks deeper into a slow server, that's when they discover the problem. What's worrisome is that in almost all of the breaches, it's actually IT noticing the slow performance of the server that's catching the attack, not the security software. What scares me is the patient, persistent attackers, because I will tell you, in all those cases, if the attackers were content with one hundred records, they probably wouldn't have been caught. There are some attackers out there that *do not get caught.*

The two primary reasons why any criminal in both the physical world and cyberspace *does* get caught are either because they get greedy—which is usually the case in cybercrime—or they get cocky and start bragging and telling other people about it. It's the attackers who lay low that I worry about, because they're the ones who can be in networks for five or ten years without detection.

That's what happened with Stuxnet. Stuxnet was an attack against a PLC (programmable logic controller) on the only nuclear reactor in Iran. A PLC is a specialized device that runs ICS (industry control systems). PLCs are built for high reliability and are critical to the operation of these control systems that run electrical, nuclear, oil, and gas. The PLC is the brains of critical infrastructure, so if it gets compromised, it could have a high impact on the reliability and safety of those devices.

The Stuxnet attack had the ability to overspin, burn out, and melt down the reactor. No one understood how the attack happened because that system was not connected to the internet. It turned out that the malicious code was installed at manufacturing time. There were over thirty other PLCs that had the same malware that was never activated. These systems that work with nuclear reactors go through extensive testing, verification, and validation; yet, they weren't able to detect the malicious code, and it was in that system for over nine years before it was activated. So if malware could be in a PLC for a nuclear reactor for nine years, I think we can agree it could easily be in networks for a long time. So far, the longest I've ever seen was a client—a large bank—and before they hired us to help them, they were compromised for eleven and a half years without ever detecting the attack.

GET CYBER HEALTHY

Let's go back to how I said that IT performance, not the security software, is what's catching attackers. Any security vendor you talk to, or any security vendor that hears this, freaks out about that statement because they will tell you they caught the attack.

Here's the problem: The security software we have today generates way too many false positives. Your security software is alerting you to a thousand things, saying, "Hey, this is an attack" when only ten of them are actual attacks. So yes, if we want to be fair, the security software is detecting the attack, but there's so much noise that the company can't find it, react, and respond to it.

Out of all those daily alerts, your security team can probably only handle about fifty attacks. So the statistical probability of actually catching a real attack is pretty low. It would be better if your security software only generated the number of highest priority attacks that your security team can handle. However, security vendors don't like doing that because that means they're missing some attacks. They'll argue such reduced capability means they'll only be catching 20 percent of the attacks, but that would be a whole heck of a lot better than the zero they are really detecting now.

But that mind shift hasn't occurred yet with most companies and vendors, which is why we're still in the situation that we are today. I'll go to companies where their security operations center (SOC) is generating a ton of alerts, and they're either not tuned correctly, or the team isn't big enough to be able to handle all of them. That's where we have to be careful and start to change our approach.

In terms of compromise, the two big things that you want to focus on as a business or an executive from a high-level strategic standpoint are:

1. How do you verify or authenticate people who are coming into the network (even more important when supporting a remote workforce)?
2. How do you detect attacks in a timely manner?

REMOTE WORK

As I am writing this, we are in the midst of the COVID-19 pandemic, and almost overnight, every organization had to move their entire office to a virtual office and support a remote workforce. By the time this book goes to print, hopefully businesses will start to reopen, but still, supporting a remote workforce will be the new normal. The most important thing to remember in working remotely is the importance of protecting and securing the endpoint computer. In a normal office environment, the company provides you with a computer, updates it, and puts you on a network that has a lot of different security devices that filter and catch attacks. With a remote workforce, the user is on a home network (which is probably open wireless) and potentially using an outdated system that is directly connected to the internet. With a remote workforce, you can achieve a similar level of functionality, but without the proper attention, security is significantly lacking. Therefore, the following is a quick checklist to help support cybersecurity for a remote workforce:

- Utilize an updated operating system that is fully patched
- Secure and lock down wireless communication
- Utilize two-factor authentication
- Install endpoint security on all desktops
- Turn on security features in software
- Use a second computer for checking email and surfing the web

Back to looking at your business overall ... I do a lot of expert witness work at both the individual and business levels, so I'll tell you that fraudulent wire transfers are a major problem. I run across this at least once or twice a week, where a business will get phished.

One of the cases concerned wire transfers between a new CFO and the company's CEO. The CEO would send the CFO emails constantly, asking the financial officer to make transfers. In the beginning, the CFO would go back and verify each one, and the CEO every time would confirm.

One day, an email came in from the CEO that said, "I need you to wire money to this account."

It seemed a little off because the amount was a little larger than normal, the account was a new one that they've never used before, and it was overseas. So there were some red flags. The CFO forwarded it to the CEO's email and said, "Can you please confirm that this is a good, valid transfer?" Within about five minutes, a reply came back from the CEO saying, "Yes, this is valid. I need this done as soon as possible. Please take care of it today."

We'll make a long story short. The attacker actually compromised the CEO's email, so the attacker's response came from the CEO's account, but the authentication failed (see below), and then the CFO transferred the money. The next problem was there wasn't any alerting in place (also see below), so by the time they figured out there was a fraudulent transfer, it was almost five days later. Typically in these cases—and check with your bank—based on my experience as an expert witness, if you can catch it within the first twenty-four hours, and sometimes forty-eight hours with international transfers, you can reverse it or change it. After that, the money is long gone. And then, once again, the company legitimately did the transfer, so typically the company is liable, not the bank. We saw a similar situation in Chapter 1, where a couple lost their new home and their purchase price in a similar phishing scheme.

I'll introduce you to two solutions now that will help put you a bit more at ease, but we will talk about these throughout the book. You want to look for proper authentication. You want to use something we call "two-factor authentication," sometimes abbreviated 2FA. With this system, every time you try to log in or access information, you must go through a second step; you don't simply use a password because, let's face it, passwords can be spoofed, phished, and compromised. You might see this with some of your banking or e-commerce sites. You log in and give your password. After you hit enter, you then get a text, or you have a token that generates a unique one-time password, which you would enter into the system, and that would authenticate you. We have still seen cases where attackers can compromise the 2FA with cell-phone hacks and other devious workarounds, so it's not perfect, but it's much harder. With some awareness, and never sharing the two-factor authentication with anyone (since banks and companies will never ask for it), it can go a long way toward making transactions and communications secure. You should always use two-factor authentication for any exchanges that are important or critical to your business.

Second, you want to enable alert notifications. One safeguard I practice with both my work and my business bank accounts:

Any time there's a transfer, I receive a text alert, which I can then at least glance at. Now yes, if you're with a really big company, it might not be practical for you or one of your employees to look at every alert, but somebody should be doing a validation to ask, "Is this real?"

The rule is that there should always be a threshold parameter. If you're doing a wire transfer to a new entity or a new customer, or if an account is altered or modified, that should always be treated very, very carefully. Most of my clients do lots of EFTs, bank transfers, etc., but they are usually set amounts each month, and to known accounts they've used in the past. Those are relatively safe and don't worry me much. If all of a sudden one of the recipients changes their account, they provide you with a new account, or there's a one-off transaction, those are the ones we have to pay very special attention to. It's the anomalies that you want to be aware of. The trick with cybersecurity is that you want to understand and identify the patterns so it makes it easier to spot the anomalies. It does not matter how good or advanced the adversary is, they are always going to make changes or do something different, and being alert to that is the key to timely detection.

YOUR DIGITAL INFORMATION IS VULNERABLE, SO *YOU* ARE VULNERABLE

When I do keynotes, I always ask the audience, "What are the two most dangerous applications on the planet today?"

A while back, a popular answer was *Angry Birds*. More recently, it's been Facebook. Regardless, what it really comes down to is the two most dangerous applications on the planet are email clients and web browsers. Those are the sources of almost all evil. Currently, the number one method to gain access is phishing. At any moment in time, you are one click away from being compromised. Any time you get an email, you need to be super careful because emails can look legitimate. We talked previously about how attackers can easily compromise mail servers so you can receive a seemingly legitimate email. You can even reply or forward that email to the legitimate entity, and the attacker would reply on their behalf because the account is compromised, so there's nothing visible that you can immediately pick up on.

Typically, there are three things that happen to help you spot a phishing attack:

1. It comes at an unusual time.
2. There's an urgency.
3. The email elicits an emotional response from you.

There are also two general types of phishing attacks; we'll call one a positive-response, and the other a negative-response email. One positive case I was recently involved with concerned a client who went home from work around 9:00 PM and received an email from her boss. It said, "Congratulations, you've been doing so good this year. We just had a meeting today and we found $3 million of excess funds. We want to increase your budget for the year, so we can actually be more aggressive in X, Y, and Z. Please review the attached spreadsheet to see if you agree with the new allocation of your increase in budget of $3 million. Because this is highly urgent, if we don't hear back within two hours, we will give the money to somebody else."

Notice how it came in so late, the urgency of time, and the emotional inclusion of getting rewarded. What do most people do? They click on it. Here's the funny thing. When I talk to people after the fact, most of them say, "Eric, something felt wrong. Something felt off, and when I clicked on it I was like, 'Ah, I probably shouldn't be doing this.'" However, they clicked on it anyway. That's the problem.

Now let's step back. What is the probability that if you were going to get a budget increase, your boss would email you at 9:00 PM, and then what's the probability that if you didn't reply back in two hours, you would really lose the money? If your employer is going to increase your budget, they're going to increase your budget. Whether you replied tonight or tomorrow, they're not going to take it away. But it triggers that emotional response in people, and that's what gets them most of the time, even if their commonsense alarm bells go off.

In this case, the best solution is to use out-of-band verification. If that was a real email, your boss would be working and waiting for a response. Therefore, pick up your phone and text or call your boss to verify the email is legitimate. In almost all of the cases of phishing attacks and fraudulent wire transfers, the problem could have been avoided or mitigated if the person did out-of-band verification. It is important to understand high-risk situations and take the two minutes to verify and validate.

I see the negative one a lot more—and it can be downright scary if you don't know what you're dealing with. You get an email in your box that says, "I've been watching you. I broke into your computer several months ago, and I've turned on your camera and been recording all of your activities. You've been a bad person. You have gone to some very inappropriate sites and have done some inappropriate things that I think if your coworkers, your family, and your spouse found out about, they wouldn't be very happy. I don't want to start trouble, but I should be compensated for my work. Therefore, if you give me $500, I will uninstall the software, and this problem will go away. If you don't, I will forward this to all your family members."

The scary thing is evidently a lot of people must do inappropriate things because many people either click on it, or I get a lot of forwards from people asking if it's real. Once again, that's a negative emotional response, but with the same strategy. It's saying you need to reply in a certain time period about a certain thing, and it's asking for a reasonable number. That's the big shift we've seen, as mentioned earlier. Attackers of ten years ago would ask for $10 million, which very few people would pay. Now they're realizing it's a numbers game. Instead of trying to get $10 million from one person, they get $500 from 10,000 people.

The other thing you need to be careful of is clicking on ad links. Most people don't realize that in social media, those ads that appear are not actually verified, validated, or checked by that social media channel. Anyone who's paying money can post an ad. A lot of those ads are fraudulent and criminal, and when you click on them, they either infect your computer or steal your money. The two things you really have to be disciplined about are clicking on emails and going to websites. If I told you to never click on attachments in emails, never click on links, and never surf websites, your cyber world would be a very safe place, but that wouldn't be practical. People need to do those things in order to do their jobs.

What I'll tell you is that most of the attacks that we see out there today are for Windows operating systems. It's not that Windows is more vulnerable than Linux or Mac. Our company tracks the number of vulnerabilities per operating system, and it's all within 1 percent of each other. The number of vulnerabilities in Windows versus Mac versus Linux is all about the same. The reason why Windows is targeted is that 90 percent of all computers on the internet are using a Windows operating system.

So you have two options. The first: If I'm looking at an attachment or a link that I'm not sure about and/or I'm surfing the web, I always do it from my iPhone or iPad. Now, once again, the probability of compromise is very low because most of the malware is written for Windows, and the iPad that I use doesn't have anything on it. I just use it for basic checking, verifying, and validating. It's a smaller footprint than my laptop, so it's easy when I'm remote and going between client sites. Even if something was malicious for iPads, which is extremely rare, the impact would be minimal because I have so little data on my tablet.

The second trick, which is also what I do in my office, is that I have two computers. I use one computer only for checking email and surfing the web, and it has no sensitive data on it. My second computer has all my sensitive data, but it's not directly connected to the internet. Now, if I'm checking email or surfing the web, and I make a mistake and get compromised, it only affects my one system that doesn't contain any critical data.

If for some reason you are transferring a large number of sensitive documents to the office, you would want to set up a VPN (virtual private network) that is directly connected to your office with 2FA. Transferring the file is relatively low risk, while surfing the web or checking email is high risk. I have several CEOs that I work with who actually have three or four computers. Based on the risk and exposure of what they are doing, the $1,000 for an additional high-end computer is well worth the extra level of protection they gain by keeping their computers separate. The bottom line is to understand the functionality you need and put measures in place to minimize or reduce your overall risk.

CHAPTER 2 REVIEW

This chapter has been just a quick illustration offering some practical ways that you can protect and secure your information online, and we'll get into more later. The point is that we live in a digital world. Most of your time is spent in cyberspace, so you need to recognize that you are a target, and cybersecurity is your responsibility.

You will be hacked at some point if you haven't been already. You need to change your mindset so you can start being better prepared, because if you have a computer, and it's accessible from the internet, it's hackable. Google yourself so you know what information is already out there about you. Maintain a social media presence so you can monitor for bogus accounts.

Consider getting throwaway laptops if you're going to countries with known hackers and you have valuable information they might want.

Cybersecurity is less about getting robbed and more like getting sick. You will get hacked just like you will get sick, but there are things you can do to minimize the frequency and level at which it happens. If your IT or cybersecurity team at your business says you aren't getting attacked, most likely they're just not detecting it.

Use two-factor authentication and banking alerts whenever possible as actionable ways to start to get cyber healthy. The key to timely detection is to understand and identify the patterns so it makes it easier to spot the anomalies. Email and web browsers are the most dangerous hot spots for breaches, as the number one method to gain access to your data these days is through phishing. You can also use out-of-band verification, meaning using another method outside of email, like calling or texting, to validate and verify. Macs are still vulnerable, but most malware is written for Windows because 90 percent of the world's computers run on a Windows operating system.

THREE

THE HACKERS ARE HERE

Hacking is big business. Organizations are compromised on a regular basis, and individuals' identities and bank accounts are constantly stolen. The problem is that they are going undetected. Whether we like it or not, we need to recognize: The hackers are here.

On average, businesses are compromised for over three years, and individuals are compromised for nineteen months before they ever detect that it's occurring. People are attacked both personally and in their business, but they just don't realize it until it's too late.

As I write this, there was an attack last week in California where a power grid went down for seven hours. It was due to a cyberattack, but the company said it was an infrastructure problem in order to keep people calm and deflect attention. It's incredibly misleading and dangerous, and the reason we don't believe any of this is real or possible is because today's attackers are stealthy, targeted, and data focused.

THE CYBER COVER-UP

Hackers try to get into your network or on your computer and monitor and extract information whenever they want and not get caught. The phrase we commonly use for this is "data theft." As I said, I think that's the wrong phrase because "theft" means somebody stole something that belonged to you and you no longer have it. If your car was stolen from your driveway, you would know it—it would be gone.

However, that's not what's happening in these cases. Hackers aren't taking a company's data or individual information and deleting it off the box so when you go to access it, it's not there. They're just copying the information. Your data is still there, your tax returns are still there, your bank information is still there.

Unless you're doing proper monitoring of your networks and your computer, very often you're going to miss these attacks. If a major hotel chain has hundreds of people working on security, and spends millions of dollars a year on security devices, and they weren't able to detect an attack for over three years, that's a pretty good indicator that their current security strategy is not working.

The problem is not only that companies aren't detecting it, but organizations are afraid of the unknown. If you hear that there was a car accident because a light malfunctioned or there was a power failure, most people might be a little concerned, but they understand it happens and accept it. However, if you were told that there was a car accident because of a cyberattack that took control of the safety system and went undetected, it would create a much bigger concern because people are afraid of what they don't understand.

Therefore, there is an overarching strategy to downplay, cover up, or not admit to a cyberattack unless you absolutely, positively have to. One trend that I notice: Whenever a cyber-attack takes place, the first thing that always occurs is that the company denies it and pretends that it did not happen. Once they are forced to admit that they were attacked, they underreport the numbers and downplay it, making it sound like a much smaller attack than it was. Over the course of several weeks or several months, the truth comes out, and we learn that the attack was a lot worse than what was originally reported.

This is important to remember for two reasons. First, if your company is compromised, be careful of what you report and make sure you have the facts. Report what you know, but include proper caveats if you are not certain what happened. It is also critical to immediately consult attorneys, since there are both current laws and new laws being passed that require timely disclosure. It is important to note that in many states and countries, there are not only disclosure laws, but there are time periods in which disclosures must be made. For example, in some cases, regulations state "if an entity believes that they might be compromised" you must notify anyone who might be impacted within a certain time period, such as forty-eight hours. The important point to highlight is that even if a company "believes" they might be compromised but they cannot prove it, the burden still falls on them to do proper disclosure.

Second, when you hear about a cyberattack that could impact you or your business, remember that what you hear is probably underplaying the truth, and it is probably worse than what was originally reported. Therefore, it is important to be proactive and take action to protect and secure your own information. If you are not sure, assume that your information has been compromised. It is better to protect your business and have it turn out that your data was not compromised than to assume that you are safe, ignore the warnings, and find out later that your data was impacted, and now you have a much bigger issue to deal with.

PREVENTION PLUS DETECTION

As mentioned, if you use a computer, and it's connected to the internet, it is hackable. Of course, it all depends on the level of effort that's required, but if it's accessible, somebody can get in. It's sort of the same premise as a house. You could put locks and alarm systems on your home, but if somebody really wanted to get into that house, and they put enough energy and effort into it, they would still be able to. Locks don't necessarily keep everyone out; they just make it more difficult, which is why you also want to monitor.

With cybersecurity, that means there are two important pieces. Prevention is ideal, but detection is a must. The problem with most organizations today is they focus solely on prevention. They try to prevent attacks, whether by putting in firewalls or with endpoint security on all work computers.

You can't prevent all attacks and here's why: Preventive devices are going to block things that are 100 percent bad 100 percent of the time. So, what if you have a situation where you have some activity that is only bad 80 percent of the time? Well, your preventive security can't block it, because things that are 80 percent bad are still 20 percent good. As an example, consider the content of messages and the ability of automated security programs to detect attacks. An email that says "Please read," which includes an attachment, is malicious 80 percent of the time—it's a common method for attackers to trick people into clicking on attachments. However, there are some cases when a legitimate friend or business associate might type "Please read," and include a valid attachment.

If you started blocking things that are 80 percent malicious, you would also be blocking 20 percent of legitimate communication and items, which would be completely and totally unacceptable to a business. So in cases where something is 80 percent bad, the preventive security needs to allow it through for the 20 percent good. That's why preventive solutions are never 100 percent. They're good, but there are always going to be attacks that sneak in, and that's why detection is a must.

Detection is more difficult because it requires monitoring—somebody to take some action. As mentioned in the previous chapter, most of the security devices are generating way too many alerts. Just think about your work computer. How many times does your antivirus software pop up and say, "Could be a problem, could be suspicious, could be malware." Maybe the first or second time you see it you'll actually do something. After that, you just start ignoring it because of the false alarms.

That's the problem that we have today in cybersecurity. The security devices in place are technically detecting the attacks. After many of these big breaches, you often hear security vendors come out and say, "Even though the company didn't listen or respond, we did our job. Our software was able to detect the attacks, so you should buy our technology." That is like me standing outside of your apartment building and saying, "I'm going to tell you when you're going to be robbed." And every time somebody comes into the building, I text you, "You're being robbed." Throughout the day, among hundreds of legitimate people who are entering the building, the robber does come in and burglarize your apartment, and I say, "See? I told you when you were being robbed. I detected the robber." If I alerted you to every single person who went into your building, and one of them happened to be a robber, that's neither helpful nor beneficial. As mentioned in the previous chapter: The solution to that is to tune your security software to respond to the highest priority attacks—the ones you can do something about. The main way of doing this is to properly tune any detection to be more specific and to eliminate as many false positives as possible. The trick is to focus on the quality of the alerts, not the quantity. Using the previous email example, instead of flagging any email with an attachment (or link) that says "Please read," it is better to tune the system to only flag emails that say "Please read," that come from an unfamiliar address from which the company has never received any emails in the past. Also, recognize that the most important thing at a company is critical data; therefore, any attacks that focus on accessing or exfiltrating critical data would be a high priority.

HOW AN ATTACK WORKS

If you're going to hack or break into an organization, you need a target. There are two general targets: servers and individuals.

Target One: Server

A server is a computer on the internet that has an address. Remember earlier in the book I mentioned my hotel room filled with gold? You only need three pieces of information: the hotel, the room number, and some weakness or vulnerability that would allow you to get into my room.

That's pretty much all hacking is. If you're going to hack a server, you need to go in and identify systems that are visible. You need to identify ways into those systems and vulnerabilities that could be exploited. Now, this is not a technical hacking book, and I'm not going to turn you into a professional hacker like one of my earlier books, *Hackers Beware*. However, as a business leader and/or executive, you need to be able to ask the right questions in order to protect your organization.

One of my favorite quotes is: "Smart people know the right answers. Brilliant people ask the right questions." Things you want to start looking at in your organization and asking about are:

What is our critical data?

Where is it located and who has access to it?

What systems or servers contain that data?

What is the visibility of those devices in order for someone to break in?

If you look at all the major breaches that targeted *servers* (which are most of them), like the Marriott or the US Office of Personnel Management (OPM), they had two things in common. First, they all had a server that was visible from the internet that had open services running. They were missing patches. When a product is released, and an operating system vendor realizes there are vulnerabilities in it, they will release a patch for it to protect the system. New vulnerabilities are discovered all the time, hence why you have software updates. So when Microsoft or Apple discovers a vulnerability in their product, they're going to release a patch. Essentially, an unpatched system is a vendor telling the world, "Listen, there's a vulnerability in our system, and here's how you fix it." If you don't apply the fix, you are in an exposed state, so patching systems is critical. Make sure your patches are up to date.

Secondly, the major breaches also contained critical data that wasn't properly encrypted. Any system that's directly accessible from the internet should never contain critical data. That's simple to figure out. There should always be control measures in

THE IMPACT OF THE ATTACK

The US Office of Personnel Management (OPM) breach compromised twenty-two million records, which ironically today is considered a small number. If this breach happened now, many of the news channels that I work with would not feel the story was significant enough to cover. The problem with this mentality is: What if you are one of the people whose record was compromised? This idea that the theft of twenty-two million records is not significant enough for people to be notified by the news is absurd, and it shows the lack of attention to critical issues like cybersecurity.

The other very important difference concerning the OPM breach is the type of information that was compromised. Typically, when credit card data is taken, it is inconvenient, but a credit card can be changed, so there is minimal long-term impact. However, the OPM database contained information on clearances—which included social security numbers and other key data that is used to determine someone's identity—that is very difficult, if not impossible, for someone to change. What made the OPM breach so significant is we now have a large group of people who will have to live the rest of their lives with their social security numbers compromised and their identities stolen.

place to protect that data. You want to make sure that your data is properly encrypted.

Here's the critical part: You want to make sure that the cryptographic key that's used to encrypt your data is stored on a separate server. In almost all the major breaches, the data was encrypted, but the key was stored with the data. That would be like taping the key to the outside of the hotel door of the room filled with gold so that you don't lose the key. You're probably laughing because that defeats the purpose of having a key in the first place. For the key to work, it has to be protected, secured, and separated from the door. Yet that's what most people are doing in cyberspace. One of the things I always say: In the real world, we tend to have common sense, but when moved to cyberspace, all common sense is lost. (We'll talk a bit more about data encryption in Chapter 5.)

Just to show you how bad the problem is, during one of the cases we worked on, the cryptographic key to decrypt the sensitive data on the server was actually stored in the log file in plain text. Generally, a log file is a text file that records all the events and activities that take place on a server. The client's rationale was that the log file has hundreds of thousands of entries, so having the key appear once or twice is not that big a deal because it's not like anyone's going to find it. I don't agree with that. One of my staff members examined the log file, and I asked how many times that plain text key appeared. He smiled and said, "Ninety-three." Well, that wasn't too bad, but he started laughing and said, "Ninety-three thousand times within a twenty-four-hour period." Now, you don't have to be a security professional and have super ninja skills to find a piece of information that appears 93,000 times within a twenty-four-hour period.

To recap: If you have a visible server that has missing patches and contains critical data that's not properly encrypted, it's going to put your organization at risk. That's it—that's all the attacker needs.

Target 2: The Individual User

Target two is the user, the individual. It turns out, that's one of the easiest targets out there because many people trust and believe in email. As I've explained, the number one attack against individuals is what we call "phishing," which is essentially where somebody sends you a legitimate-looking email with the hope that you click on it. When you do click on the link or open the attachment, it infects your system. One click is all it takes for you to be compromised.

You're going to get a legitimate-looking email from your boss or it might be a legitimate-looking email from Amazon, the government or the IRS, or many other entities that are out there. One thing the attacker is really good at is taking advantage of opportunity.

As mentioned a bit earlier, as I'm writing this book, we are in the midst of the COVID-19 pandemic, where everything is shutting down, opening up, shutting down . . . It's something I've never seen before in my life, and hopefully by the time you're reading this, it will be behind us. But right now, we've seen so many people get their computers infected because the attackers are sending out emails that look like they're coming from your insurance company, your doctors, or even the center for disease control, touting critical, though fake, information about

COVID-19 that you need to protect and secure your family—all of it encouraging you to click.

This is a new territory, and it's scary. But unfortunately, we have to remember the adversaries have no morals or ethics. They're going to take advantage of every opportunity they can. As I said before, I recommend that whenever you're checking email, do the first pass on a non-Windows device and a simpler device such as an iPhone or an iPad. While a MacBook could be used, not because it is more secure, but currently it is less prone to attacks since it has a smaller user base, it is important to remember that it is a full-blown operating system. The more complex an OS, such as Windows or Mac, the more vulnerabilities that are present.

Tablets and smartphones, like iPhones, iPads, and Android devices, are simpler devices. They don't have as much complexity. They don't allow as much embedded code to run on them. And very few (I'm not saying any) pieces of malicious code actually run on these devices.

These are just a few of the things you can do to minimize your exposure and lower your risk. We'll get into more ways to protect yourself later in this book . . . and what to do if it becomes a real threat for you.

TAKING STOCK OF CYBERATTACKS

Cyberattacks are a fact of life these days, but people downplay them, as was discussed at the beginning of the chapter. Companies, organizations, and entities recognize that when we say "cyberattack," people get very frightened because the cyberat-tacker is an unknown entity and it's unknown territory. There-fore, they would prefer to just minimize it, dismiss it, or pretend it didn't happen.

A perfect example was during the Iowa caucus of the 2020 Democratic primaries. The day of voting was a total disaster because the apps didn't work. They crashed, they failed, and they weren't able to report the results that day like they were expected to. It took a number of days to get accurate results. The interesting thing about that is, as soon as it happened, within fifteen minutes of the news that there was a problem, the offi-cials immediately came out and said, "We can guarantee it's not a cyberattack. It was just a testing functionality issue."

I was scratching my head, noticing two important things. First, how quickly they were able to come out and overtly say it wasn't a cyberattack. Why did they need to do that? They could have told the truth, which is that they had no idea at that point what caused the problems. But because of the uncertainty around cybersecurity and cyberattacks, they felt the need to make that clear statement right out of the gate. The second thing is that I knew they couldn't definitively say it *wasn't* a cyberattack anyway. If they didn't do a full forensic analysis, and they weren't able to discover the root-cause problem, then how could they accurately say that it wasn't a cyberattack? And that's the real-ity of these situations. Many organizations believe that if they ignore the problem it will go away; however, the reality is that if you don't recognize the nature of a cyberattack, an organization could be compromised for a long period of time. For example, a Virginia-based health system was compromised for 5,605 days before it was discovered.

In truth, when cyberattacks may not be recognized for days, weeks, months, or years, an entity cannot dismiss that possibility within fifteen or twenty minutes, or even two to three hours. And I will tell you with firsthand knowledge: Some of the electrical outages that we have had in this country, and some of the functional impacts that financial institutions claim that they quickly wrote off as other problems, issues, or misconfigurations, were actually cyberattacks. But for some reason, people feel the need to cover up and not accept the reality of the situation.

Remember, in cybersecurity you cannot prevent all attacks. Prevention is ideal; detection is a must. The right areas to focus on are access to critical data and to monitor for unusual activity. Most organizations focus their cybersecurity monitoring on inbound traffic, but where does most of the damage occur? Outbound traffic. If data is compromised, stolen, or exfiltrated from a company, it all occurs on the outbound path. Monitoring outbound traffic is critical for catching attacks.

Security is all about timely detection and containing and controlling the damage. If you're catching attacks within twenty to thirty minutes, that is unbelievably good. I know when I talk to executives, they say, "But Eric, you're telling me that no matter what I do, no matter what I put in place, if we're a functioning business, then you're saying we are going to get attacks?" And the answer is yes.

Ask your team or whoever's overseeing your security:

1.   How many attempted attacks do we have a week?
2.   How many attacks are getting in?
3.   How quickly are we detecting those attacks?

If they can't give you a specific answer on the number of attacks and how they're detecting them, you need to change or adjust your cybersecurity techniques or strategies. The only real metric of success is how quickly you are detecting the attacks and how much damage is being done.

Now, it's also important to recognize when we're looking at this area that security and IT are very closely related. As I said, most of the companies—small, medium, and large—ultimately detected their major breaches through IT performance issues; they had servers that were operating in a non-optimal way. The memory was overloaded; the network traffic was too high; there were performance issues on the network. And when the IT team started investigating those performance issues, they were able to detect the attack—even if it was a few years down the road. So you want to urge your team to always baseline your systems, and whenever they see any deviation from a known baseline, you want them to step in.

THERE'S GOOD NEWS: CYBERSECURITY IS EASY

I know I may have pumped you full of intimidating information, but cybersecurity isn't really that hard—it's actually a pretty short list, so it's not like you need to do five hundred items to secure your server. As we've learned, if you either have servers that are visible from the internet that are vulnerable or exposed, or if you have individuals who are getting malicious activity and clicking on the link or opening the attachment, you're setting yourself up to be a target. As an executive, to protect and secure your organization, you should ask some basic questions of your tech team or IT professionals about which systems are visible from the internet. Are they all required to have that visibility? Which ones are not patched? Essentially, it is important for an organization to understand their exposure points and make sure they are protected. Many organizations that are compromised were compromised because they had servers they were not aware of, or vulnerabilities that they did not know existed. Cybersecurity is not difficult if you have proper awareness. This awareness starts with knowing what servers are on your network (asset inventory) and understanding the software that is installed and how the software is configured (configuration management).

Some of the specific questions your team should know the answer to are:

1. What is your critical data?
2. Which servers are that data located on?
3. What servers are visible from the internet?
4. Do any internet-facing servers contain critical data?
5. How many internet-facing systems are missing patches?
6. Is our critical data encrypted?
7. With encrypted data, are the keys stored on a separate server?

Some bad news is that most companies are not even fixing the easy stuff. Some balancing good news is that many of today's attackers are not terribly sophisticated because companies have been sloppy with their security.

If companies make sure that all of their servers visible from the internet are up to date, fully patched, and contain no critical data, and if your cryptographic keys are stored on a separate, secure server, none of the types of major breaches that we've seen over the last years would happen. By just focusing on these key core areas, you can make a significant difference in your overall security.

Now, what gets a little more difficult is the user angle, because companies generally need email and web surfing in order to function. Most companies in the past have typically said, "If anything impacts our business, we're not going to do it." But today we live in an age where there has to be a trade-off. Remember the risk-versus-rewards equation we examined earlier? The analogy I often give executives is there is going to be a little pain. Do you want to be shot in the leg or shot in the chest? Most executives say, "I don't want to be shot." And I reply, "I agree, but you're going to be."

Well, if it has to happen, you'd rather be shot in the leg. What I'm referring to is this: Does your company really need to allow email messages from unknown entities that contain attachments and embedded links?

I've done this analysis for companies, and what it typically comes down to is if they allow attachments and links from unknown entities, it often costs a medium- to large-sized company five to six million dollars a year. If you block attachments and embedded links from unknown entities, it might cost you

$200K in some job performance, because there might be a little impact where certain job and processes aren't working at an optimal level. This cost is based on productivity impact to the organization. For example, if someone needs to exchange files, and instead of using email, they now have to use a separate piece of software to exchange the files, this could increase the time it takes to perform a job and cost the company more money.

So the question is, do you want to lose $200K or do you want to lose $6 million? That's really what we're talking about. You need your security team to provide those numbers to you. One of the biggest failures in cybersecurity is that we're not providing accurate figures so executives can make the correct decision they need to protect and secure their environment. If you're not getting those numbers, ask for them: "If we continue to operate the way that we are, what is the impact on the business? And if we implement your suggestions of blocking attachments and embedded links, what would the impact be?" And then base your decisions on factual figures so you can arrive at the correct solution.

To get back to the good news: As I said, this is not hard and there are actionable things you can do. I've provided a few of those to you in this chapter, but as you continue to read this book, I will provide more and more solutions for you to protect and secure your organization.

CHAPTER 3 REVIEW

Individuals and organizations are hacked on a regular basis. The problem is, because there's usually no visible sign, people don't realize it. If your company is compromised, be careful of what you report and make sure you have the facts. When you learn about a cyberattack that could impact you or your business, remember that what you hear is probably underplaying the truth, and it is probably worse than what was originally reported.

With cybersecurity, prevention is ideal, but detection is a must. You want to start asking: What is our critical data? Where is it located and who has access to it? What systems or servers contain that data? What is the visibility of those devices in order for someone to break in?

Update your systems with patches. Definitely make sure the cryptographic key that's used to encrypt your data is stored on a separate server. Hackers will most often try to target your server with an encryption key if it's accessible to them, or they may enter via email phishing.

Cyberattacks happen on a regular basis. Monitoring outbound traffic, not just inbound traffic, is critical for catching attacks. Ask your team or whoever's overseeing your security: How many attempted attacks do we have a week? How many attacks are getting in? How quickly are we detecting those attacks? The only real metric of success is how quickly you are detecting attacks and how much damage is being done.

If you have vulnerable or exposed servers that are visible from the internet, or you have individuals who are getting malicious activity and clicking on a link or opening an attachment, you're going to become a target and get broken into. You need your security team to provide performance numbers so you can make the best decisions to support your business.

FOUR

MOBILE WEAKNESSES

WE'RE ALL WALKING AROUND WITH A HACKABLE DEVICE

Everyone talks about the importance of personal privacy. However, the second you bought a cell phone and started carrying it around with you, the privacy train left the station and is never going to return. Your phone is a hackable computer that's always on you. Anything you record or put on your device is going to become public. Period. Everywhere you go and everything you do is being recorded by that little device. Your activity, your location, everything you search—it's all being recorded on any mobile device including your cell phone and tablets. Yes, even if you have certain settings turned off. Because here's the issue: Many free apps—apps for coffee shops, for running, for other things like that, which we'll talk about later in the chapter—need to be able to utilize certain features of your phone, like location services, in order to work.

Your life is automatically recorded just by having apps or performing certain activities. Most of us would violently object to a chip in our arm tracking our every movement, yet that cell phone that's with us, and never more than two feet away, is doing just that and recording a lot of personal information.

For example, in military operations where they've wanted to identify and remove hostile entities, they've actually utilized that person's cell phone in order to target them with what we like to call a "special gift" or a guided missile or drone. So we're seeing cell phones used not only as nice, convenient devices, but also in warfare and military operations in order to track and identify where somebody is and potentially use that information to take that person out.

As a commissioner on cybersecurity for the forty-fourth president of the United States, I got involved when President Obama wanted to be able to carry a cell phone with him. It was ultimately decided that they would give him a BlackBerry because a Black-Berry was a lot more secure than other devices. I remember sitting in those meetings where we had to ask questions like: *What if somebody gets a hold of that cell phone? What if somebody takes it? Think of all that critical information. How do we make sure we have encryption and protection and all those measures in place?*

I would just laugh and tell them they were missing the whole point. With all of the Secret Service around the president, what is the probability that the president is going to lose his phone or that somebody's going to steal his phone? Probably zero. Because if he left it on a table, somebody attending to him would pick it up, and nobody is going to get close enough to be able to steal the president's cell phone. The whole focus on encryption of the data was missing the point. The main concern with the president having a cell phone is the ability for someone to track his specific location, 24/7, not necessarily the data on it. This is an important lesson in cybersecurity—to make sure you are fixing the right problem.

We must rethink our mobile devices and mobile footprints when it comes to cybersecurity because there's often a lot more at risk than we realize. And as a former professional hacker at the CIA, I can tell you, as mentioned earlier, that any device that has internet access is hackable and that includes your phone or tablet.

In most cases, we are our own worst enemies. Our phones can be broken into not because of their default install, but because of all of the apps that we add to those devices. Think about how many apps you have on your phone that you haven't used in two, three, or four months, or even years, and they're just sitting there on your device. The scary thing is that those apps, even unused, are actually causing security issues, exposure, and harm.

One of the most dangerous words when it comes to the internet and cybersecurity is the F word. I can't stand the F word. I hate the F word and I wish the F word had never, ever been created. The F word that I'm referring to is "Free."

When it comes to applications, and in this case apps for your cell phone, free is not free. What free is saying is: "We will

IS TIKTOK REALLY A SECURITY CONCERN?

Any application that you install on a mobile device that requires microphone access could potentially be used to spy on people or gather personal information about the devices that it is installed on. At the time of writing this book, there have been a lot of "concerns" that TikTok is Chinese malware; however, these claims have not been proven. While you could argue that there is a good chance it is true, similar could be said of many of the social media applications. It is also important to look at the purpose of the application and the user base. The application is not meant for business communications or use. Honestly, I would be more concerned with WhatsApp, since that is being utilized for a lot more sensitive communication than TikTok. Secondly, the user base for TikTok is mainly teenagers, posting videos. If you are a business executive and you use TikTok, may I suggest a hobby such as golf or tennis?

The bigger concern about TikTok is not the security of the app itself, but the data that is transmitted. For example, there was a "penny challenge" that was posted that involved dropping a penny between a phone charger and the wall plug, and it caused several fires because unknowing kids attempted it. While one could debate content censorship on many platforms, including TikTok, those are more of an acceptable-use issue and not a security concern.

give you this app at no cost, and in exchange, you will give us access to your critical information and critical data." It is not free, it's bartering. They are giving away your personal privacy. Seriously. With most of these free apps, the only way they will operate is if you give them access to your camera, microphone, pictures, or location.

A lot of people misunderstand the security checks of the App Store or the Android store. If there's an app that you install on your phone and it accesses your camera, pictures, microphone, or location without ever asking you or telling you, that violates the security. So that will not be allowed in the Apple or Android store. However, if the app asks you for permission, and it only has to do it once, and you say yes, then that is considered a secure app because it asked for your consent and you agreed. When most people install apps, they become so immune to this that they don't even realize it. When you install an app for the first time and you run it, it's going to ask you: "Can I access your camera? Can I access your microphone? Can I access your location?" We just click yes, yes, yes without much thought. Now that they're on your phone, they can and will access your information and use it how they please (the good news is you can do some simple checks to help, which I'll get into later in this chapter).

I only use paid apps. If an app is worthy enough to be on my phone, I would rather pay for it. You've probably always thought that if there's a free version and there's a commercial version for $3.99, who would be stupid enough to pay $3.99? My response is: Who would be stupid enough to download the free version and give away all of their privacy?

Let me break it down for you. The developer of an app is going to make money in one of two ways. One way is by giving you the app for free, but your payment is actually granting them access to your data, location, camera, and microphone so they can utilize that data for marketing purposes to be able to do targeted marketing and advertising, and make money off of the ads. The second option is they can just charge you outright for the app, a few bucks or so, and then all your data and all your information is safe and protected because they don't need that access—they made their money directly. I don't know how much your privacy is worth or how much your data is worth, but to me, it's easily worth more than $3.99, $4.99, or even $7.99.

To put it another way, the free apps will only work if you agree to give them access to your device. So the next time you download a free app and it says, "Can I access your pictures, your camera, and your microphone," say no. If you say no, the app will not work. On the contrary, if you have paid, say, three bucks for an app, it won't ask you for access to your camera and microphone because it doesn't need that access in order to function. When you're looking at apps and considering commercial or paid, always go with the paid version because, as they say, you get what you pay for, and it pays to be a lot more secure.

A PICTURE LASTS FOREVER

Anything you do or record on your phone will last forever. There is no delete button on the internet. Yes, even Snapchat. After ten seconds, sure, it removes the picture from access within the app.

But I will tell you from having done forensic work, those pictures can still be found. Looking at the application, it might seem to you that your data has been deleted, but it's actually still there. It's the same thing if you take pictures or post messages and you delete them. They might look like they're off your phone from a user standpoint, but I have done work in numerous cases where, if I can get access to a cell phone, not only can I pull up all their deleted information and deleted data, but I can also pull up exactly where they were (their locations), where they were identified, and all the other key components that ultimately people would like to keep secure and private.

At a high level, when you have data on a device and you delete it, what you're really doing is removing a pointer to that object from your device so the app can no longer access it. However, that data is still on any storage media, such as flash drives, memory cards, or a hard drive, and accessible there. It doesn't actually remove it from the hard drive. So be very careful of what you say or what you do because once you hit send, save, or post, that information will exist forever.

So now I'm going to get to the elephant in the room . . . I know the fact that I even have to give you this advice sort of boggles my mind, but we all need to learn that no matter how tempted you are, don't ever take naked pictures of yourself. I thought that was obvious until the richest man in the world—Jeff Bezos—got caught for doing exactly that. While he was still married, the founder and CEO of Amazon was nabbed for taking naked pictures of himself and sending them to his girlfriend. Not good. Now I'm not here to judge, although I don't think it's ever a good idea to send naked pictures to your *girl-friend* when you're *married*, but there are ways that you can do it more securely; never 100 percent, but more securely. You would think the richest man in the world would know that, but he didn't. So if he could make that mistake, you could, just as easily. (And don't come back to me and say, "I wouldn't be a target," because as we already established earlier in this book, you already are.)

I just find that story so funny because the way that it broke is the *National Enquirer* was trying to blackmail him, and the person there doing the blackmailing was actually named Mr. Pecker. No joke. And then the best part is the data that he was trying to blackmail Jeff Bezos with was in fact located on cloud-based services that were housed at Amazon Web Services. So his own data storage was actually being used against him.

It's a cautionary tale, even if you don't take nude pictures of yourself. Remember, any texts you send, anything you say, or anything you do, be careful because it will exist forever. This is especially important to note when it comes to texting and business. You might offhandedly text someone, and because you're not face-to-face with them, it can easily and quickly escalate and get very emotional. How many times have you texted someone and said things you normally wouldn't say or do, only to regret it later? I've seen many executives get into trouble because they start texting back and forth with coworkers, and what started off as an innocent text turns into cyber flirting and sexting or even cyber yelling very quickly and gets totally out of hand. Without even realizing it, now there is a permanent record on your cell phone, and courts have actually allowed digital evidence to be legally admissible.

I have worked on several cases in which individuals thought they deleted information off their phone, and the information was still accessible and used against them in a court of law. In one case, someone was stealing information from a company and providing it to a competitor. While the individual denied it, we were able to find pictures of the trade secrets on the phone and recover text messages in which they were sending that information to engineers of a competing company.

It is important to remember that there are no secrets in cyberspace. Any time you create any digital information, before you hit send, save, or post, ask if you want this information to last forever and if the answer is no, don't create that digital footprint.

MOBILE RISKS AND FIXES

I want to just quickly address that if you have children, make sure you talk to them about cyberbullying. Some parents get very upset with me when I say this, but I would argue that almost all children have both given and received cyberbullying without even realizing it most of the time. They're kids. They don't always think. I've seen my daughters sometimes joking around with friends, and they'll say things like, "I hate you, I can't stand you." I have to explain to them that I know they think they're joking, but online, it can easily be taken out of context. What if the principal of the school saw that you or somebody sent that to you? The sender could get in a lot of trouble. So not only do we need to be very careful about our own digital communication, but we should talk to our children about it as well.

Also, be sure to verify whom you're communicating with. Right now, there is a lot of text spamming going on. Someone might text you something like, "Eric, it's been a while. How are you doing?"

You have no idea who's at that phone number, so you go with it and say, "Hi, I am doing well."

Then they'll say, "I have this great business deal" or, "I was wondering if you can help me out with a problem. Here's a link that explains it in more detail."

I'm not saying everybody would fall for this, and the fact that you're reading this book probably means you're at a higher level of awareness, but can you think of at least five people that you know who would click on that link or open the attachment? Remember, hackers are playing a numbers game and, unfortunately, that's all it takes to compromise or break into a system.

The same goes for phone calls. The technology exists today that just by answering a phone call from somebody you don't know—just by clicking the answer button—they can drop malware onto your system. If they're not in your address book and you don't know who they are, don't pick up the call. I know many people are so curious that they'll pick up any call, but just be super careful.

As mentioned in Chapter 2, we're seeing a lot more cyber blackmailing going on, and in some cases, they're actually calling your bluff. I know of one scam that happened a few months ago that made over $200,000 by claiming they had compromising pictures or videos, or knew about questionable online behavior, and they would share that with your wife, family, or friends if you didn't transfer, say $75, to an account they give you.

The reality? It's a total scam. They don't know what's on your phone, but they figure that if they send that message to hundreds of thousands of people, a good percent probably fit that profile. There are a lot of people who probably view videos or pictures they shouldn't. There are a lot of people who probably communicate with people they shouldn't and have said things that they shouldn't. They're hoping you won't call their bluff and instead will take the bait, or rather, that even 10 percent will take it.

WHAT CAN YOU DO ABOUT THIS?

When it comes to mobile devices, I have some good news and some bad news. The good news is that almost all devices today have a lot of security built in, and that's a big improvement. If you go back ten or even five years ago, it was sort of known that iPhones were very secure and very locked down while Androids were very open and vulnerable. That's why you heard about jailbreaking (turning off the security features of the phone) on Android phones and often heard about tens of thousands of people getting infected with malware. But now, Android has made a lot of progress, and Android and Apple are very similar in how they function, and both have a very good amount of installed security.

The bad news is that security isn't typically turned on by default. We are at a point in time where mobile companies understand the importance and need for security, so they embed it into their devices for marketing and competitive purposes, but they don't feel that it's critical enough that users would accept the potential inconvenience of having it fully installed out of the box. So the security is there, but you have to go in and turn it on.

For example, if you look at passwords on some devices, the default is still the old four-digit PIN, though you can enhance it to six, eight, or even longer passwords, but you would have to turn on those features. I recommend that you spend some time turning on the security of your device. Go in and limit and block who can access your data and send messages, and make sure you have strong authentication. It is important that with any mobile device, you spend five or ten minutes under the security settings, researching the options and understanding the exposure of your device. Spending a few minutes up front can save a lot of pain and suffering from your information being compromised in the future.

Ultimately, even with all that security, it comes down to personal discipline. You need to make sure that you don't do things that could put you at risk. Don't say things you can't take back, don't take pictures that could be harmful, or install apps you don't need. And if you did install an app and you're not using it, remove it. At the beginning of each month, when you go in and you pay your bills or tend to anything that you do on a monthly basis, you should review your phone and ask yourself, "If I have not used this app in thirty days, should I uninstall it?"

Another option that you should consider, especially if you're a business owner, entrepreneur, or executive, is for you and your staff to carry two cell phones—one for personal use and one for work-related activity. Because let's face it: There are going to be things you want to do for work that are more secure, and you'll want to lock down that device. You're not going to want to put random apps or joke around or do personal things on a professional device. Carrying two devices gives you a little more mental clarity: You will keep your work device tightly locked down, secured, protected, and limited. On your personal device, maybe you visit more sites, put games on it, and have a little more liberty.

By contrast, if you have one phone for both work and personal use, and you do something personal like play a free game or visit a public website, it might put your device at risk. Now, all of your personal data, work data, and work information is exposed and vulnerable. But if you have two different devices, and you do something silly on your personal device and something happens to it, there's really no harm. You can just get a new one without the loss of a lot of sensitive data or information because your work device holds all of that critical, sensitive information.

The reason for this is the difference between a major breach and a minor breach. The difference is not the device or the computer that got broken into, but the data that was compromised. As always, think:

What is the critical data?
Where is it located?
How is it properly protected?

If you are going to store sensitive, critical, work-related or even sensitive personal information on your cell phones, you might want to look at getting some of the different encrypted apps that are out there. These applications provide an extra level of protection by encrypting information with a second form of authentication. They typically require an additional password, but that minor inconvenience is well worth the extra level of security and protection you are getting.

When it comes to installing apps, I always ask a few questions:

What is the function or benefit that I'm getting from this app?

What is the risk or exposure by installing this app?

Is the functionality that I'm getting worth that risk or exposure?

From now on, before you download an app, click on its information in the app store. It will tell you in which country the app was made, when it was created, and how many people have downloaded it.

So, first thing, if this app is brand-new and only one hundred people downloaded it, you might want to be careful. If the app's been around for ten years and millions of people have downloaded it, at least you know it has a little more public acceptance. However, that doesn't mean that all those people actually understand and track security.

I'll give you my favorite example. There is a flashlight app in the app store that has been downloaded three million times and has been around for eight years. You would think with my advice, that seems okay. However, this app is actually made in China. You can see the country where the app was developed. When you install this app, it requires that you give it access to your micro-phone, camera, location, and pictures in order for it to work. And if you refuse to provide that access, the app will not function.

Think about that for a moment. Why in the world does a flashlight app need access to all that data? The short answer is, it doesn't. So that's why you want to be so careful. Always use that risk formula to evaluate an app before downloading it:

What is the value?
What is the exposure?
And is the benefit worth that risk or exposure?

Pull Out Your Phone Right Now and Do This . . .

Remember, many cell phones are fairly secure out of the box, but you can make them more secure by turning on all the security features. You may also be your own worst enemy. Your cell phone is spying on you because you allow it to by installing all those extra apps.

Go under your settings on an iPhone (and it's similar on an Android) and under General, go into Privacy, and then you'll see different tabs. You'll see one for camera, one for micro-phone, one for pictures, and one for location services, among other things. Click on each and it will show you all the apps that you've inadvertently granted access to, without even realizing it. Shocking right?

Right before you go to sleep, go under cellular data usage, and you'll see that it tracks the usage—the amount of data that has been uploaded and downloaded from your phone. Now, scroll through all the apps to the very bottom and you'll see a "reset data statistic" button. Click it. If you then go back to the top, you'll see that the amount of uploaded and downloaded data is going to be zero in both cases. Now go to sleep, sleep well, and when you wake up, go back and look at the stats—it's not even close to zero. Now remember, you've just been sleeping for six, seven, eight hours, and you have not used your phone at all. But guess what? It was busy while you were sleeping. For most people, it's a couple hundred megabytes and I've even seen up to seven hundred or eight hundred megs of information uploaded and downloaded without you realizing it. So if you don't think those apps are spying on you and putting you at risk and putting you at exposure, you need to think again.

Before you move on to the next chapter, spend some time going into your phone and tablet and going through your apps. I am very diligent about what apps I have on my phone, and if I don't use an app, I remove it as soon as I realize that it's not being utilized.

CHAPTER 4 REVIEW

Your phone is a hackable computer that goes everywhere with you. Free apps are not free, and you are exchanging your privacy and data to use them. You'll pay for it one way or another, so it is worth buying the paid version of an app to help protect your data.

Once data exists, it never really gets deleted, and any texts you send, anything you say, or anything you do will exist forever—so be careful! Be wary about unknown texts or phone numbers—they might be malicious. Almost all devices today have a lot of security built in, but those features aren't turned on by default. Since cybersecurity is your responsibility, you'll need to turn those on. Uninstall apps you're not using.

Consider carrying one cell phone for work and one for personal use. Consider downloading a data encryption app as they provide an extra level of protection by encrypting information with a second form of authentication.

FIVE

YOUR LIFE, HANGING IN THE CLOUD

Whether you're a small, medium, or large business, moving to the cloud can actually provide immense benefits and value; however, as we've learned already in this book, any time you add functionality, it also means you're decreasing security. As you make decisions to potentially move to the cloud, it's important to recognize and understand the security risks and what to do about them. But first . . .

WHAT IS THE CLOUD? AND WHY DO YOU NEED IT?

At a most basic level, the cloud is a third-party hosting server that stores your data, files, etc. on your behalf. You can store your digital information using on-premise or cloud-based options. On-premise is where a company has its own data center, compared to cloud-based, in which you outsource it to a third party.

With on-premise, you own your data center—all the equipment, infrastructure, and you run everything—and typically because the data center is on your network, you can limit and control access to who can see and access the servers. In many cases, those servers are not directly accessible from the internet or from third-party networks.

With the cloud—because your data, your systems, and servers are potentially being run by a third party—that data now has more accessibility since in many cases the servers are directly connected to the internet. At a starting point, the cloud is nothing more than third parties running your systems and your servers.

When most people talk about the cloud, the companies that are looking to acquire the services are really referring to the public cloud—systems and servers that are essentially accessible from the internet. Based on accounts, access control, and other factors, they control and limit who can and can't access those systems. Amazon Web Services (AWS) or Salesforce are great examples of public cloud services that are really accessible to anyone. Yes, the systems are set up to be able to limit or control who has access to them, but technically those servers, those IP addresses (the unique address used to identify a system on the internet, similar to how a house address is used in the real world), and some interface for logging in, are publicly available. That means from an attacker perspective, if they can guess user ID, passwords, or other credentials, it could be easier to break into a public cloud-based system. That's why monitoring is so important (which we'll get to later in this chapter).

There's also another type of cloud, which is called the private cloud. The private cloud is where a third party will go in, run, set up, and configure your servers, but it's only accessible on a private network, a private connection, or via limited access.

If a company had a private cloud setup, the data center would be run by a third party, but the company would use either dedicated connections, VPNs, or other private lines to connect to the systems. With a private cloud, the systems would not be accessible from the internet and would only be accessible from the company's private network. This way, while technically anyone on that local network could get to those servers, the servers aren't visible outside of the private network, and, therefore, would have a higher degree of security and control than the public cloud services. You're getting a lot of the benefits of on-premise, but a third party is still running and controlling those systems.

The benefit of utilizing the cloud (both public and private) is you don't have to worry about infrastructure, services, software, and all those other pieces and components. Other people worry about those on your behalf, so, generally, equipment costs, replacement costs, etc., aren't as expensive or laborious, and you can have less staff. The issue is if external cloud personnel are running and managing those servers and systems, then they're usually ultimately responsible for the security, which can be a good or bad thing.

When the cloud first came out, many security professionals hated it—and I emphasize the word "hate." I remember companies' security teams would say, "We will never allow any of our services to go to the cloud. We prohibit the cloud under any circumstances. We will never, ever let anyone utilize the cloud."

I would laugh because a lot of the employees or people within an organization ended up signing up for cloud-based services on their credit cards and reimbursing those as monthly expenses, and nobody in IT or security was even aware of it. I would argue in this day and age, it's almost a guarantee that your company is already somehow on the cloud. I don't know any companies that are not utilizing some type of cloud-based services at some level, whether it's for sales, accounting, back office, email, or some component. The trick is to understand and embrace cloud use so your company can secure it, instead of trying to fight it.

I've always had a different view of security in the cloud than some of my colleagues. I always thought it was a better option, and here's why: If you're running your own data center, and your own on-premise data and security team is responsible for protecting and securing the system, and you are responsible for maintaining and staffing the security functions, then your company has the ultimate responsibility for making sure all of the systems are patched, properly configured, and locked down. Traditionally at a company, security and IT are overhead functions. They are not making money; they're dependent on the overall profitability of the company. If a company has a really good year and makes lots of money, then overhead increases. If a company has an average year, then overhead stays the same, and if a company has a really bad year, then overhead is going to shrink. This means in any given year, the size of the security team is going to potentially change. It can increase and decrease based on how well the company did. Let me ask you a question: Do the security needs of a company *really* change? No. You could argue that if the company had a bad year financially, that security could be even more important because a breach would have a greater chance of causing more harm or putting the company out of business since they have less money and lower profitability. Therefore, performing the security in-house, and not utilizing the cloud, would actually cause your security budget, resources, and focus to fluctuate, which could lead to security not being properly addressed in many organizations.

If you move your infrastructure to the cloud, these cloud-based companies essentially live or die based on security because security is a top priority to retain clients. It is a revenue-generating component. You're charging your companies for security; they can even pay for additional security; and there's a profit. It's now a revenue-generating activity, so the cloud provider is going to keep the same resources, the same folks, and because you have service level agreements (SLAs) in place, they're going to make sure that they deliver on that security. I actually think in many cases, cloud-based companies do a much better job with their overall security.

When I've seen companies run their own servers in-house, they usually have poor configuration management, and their systems aren't always patched the way they're supposed to be. With all the cloud providers—or at least the major ones that I've seen—configuration, management, and patching are always done very well and robustly because they know these are critical to their success. So yes, there are risks in moving to the cloud, but because it's now a business enabler, it tends to be done better.

However, here's the golden rule that I always tell all my clients: You can outsource IT, you can outsource security, but ultimately, at the end of the day, you can't outsource liability.

The role of security on the cloud is different than in-house security. Traditionally, security on your own servers is very tactical: Patching systems, configuring rule sets for firewalls, monitoring intrusion detection systems, updating the rule sets, and all those things needed in order for the system to work correctly so that employees can access the applications that are running on the server. When you're moving to the cloud, a third party is now responsible for all of that. Security is now much more strategic. Your own security staff is now focused on management, oversight, policies, and making sure proper SLAs are in place to verify that the third-party cloud provider is doing what they're supposed to be doing. When using the cloud, a lot of the traditional security tasks that might not have always gotten done in-house, because of limited or changing resources, are now consistently done by the cloud provider.

The problem is, you need to make sure it is being done at the proper level because if it's not, and there's a breach, your company is still liable. If your company is using a cloud-based service, and the cloud provider tells you that they are patching, and that all critical data is on servers that are not accessible from the internet, and that turns out not to be true, and there is a breach, your company suffers the harm. The media typically will say that your company had a major breach and will be liable to your customers. You could potentially take legal action against the cloud provider, but this is after the damage has been caused to your company. Therefore, it is much better to verify and validate that the cloud provider is doing what they say before a breach, instead of waiting until after the breach occurs.

You can outsource security, but you cannot outsource liability; therefore, it's up to your security team to make sure that you're getting the proper data, information, and metrics from that third party. Some of the common metrics that you want to make sure you know are:

How often is the cloud provider patching the systems?

How long does it take to apply a critical high, medium, or low patch?

Are there any patches that are never applied?

How often do they scan for vulnerability and remediate the vulnerabilities?

How often do they monitor and look for anomalies and respond to those anomalies?

Essentially, you want your own security team to ask the right questions and track the metrics to make sure it's being done correctly. You should set up your team to receive regular reports from the staff at your cloud provider, perhaps monthly or quarterly.

It's also critical that any time you're signing any contract for your business (because almost all contracts today involve data-storage security, and in many cases, the cloud at some level), you need to make sure your security people are involved with contract negotiations. If your security team isn't working very closely with contracts, and legal isn't providing or reviewing the language, that is a big gap that you'll want to fix as soon as possible.

Security is always going to be important. But consider, when you move to the cloud, you might only need one security person; you likely needed multiple security people when you stored your data in-house. I will tell you, there are some fantastic cloud providers, and there are some horrible cloud providers. You want to do your homework so you're properly protected and properly secured. There are always bargains, but one thing I've learned in my life is there are three things that you definitely don't want to go cheap on: parachutes, heart surgeons, and cloud providers. You get what you pay for.

THE TYPES OF CLOUD PROVIDERS AND HOW THEY WORK

There are three different levels of cloud services. When you talk about the cloud, you have Infrastructure as a Service, known as IAAS. You have Platform as a Service, better known as PAAS or PAS, and you have Software as a Service, known as SAAS or SAS. Generally, if you're using traditional on-premise data centers, you're basically doing everything. If you're using IAAS, they're essentially providing the infrastructure and operating systems, but you need to configure and run and install everything. If you're using PAS, they're mainly running the operating systems, and you just have to worry about the applications. If you're using SAAS, they're pretty much taking care of everything.

To help you understand this, let's look at this as "pizza as a service," and the different ways that you can have pizza for dinner. If you're doing the traditional on-premise approach, you're making the pizza at home. You're going to the store, buying the flour, cheese, and all the ingredients. You come home and you make the entire pizza. You're taking care of everything.

With IAAS, you can go to the grocery store and buy a ready-made pizza that's not yet cooked. You take it home, turn on the oven, bake it, take it out of the oven.

Now the other option, of course, is to just get your pizza delivered, ready to eat. That's PAS because now they're bringing the hot, juicy pizza to you. You don't have to worry about the ovens, the baking, or anything like that. But, because it's delivered to your house, you still need to have a dining table, and you need to provide the beverages and pour the drinks.

Your final option is to go to a pizza parlor and really have everything done for you. You can eat there, be served on their tables and plates, and they will clean up when you're done. That's SAAS.

Let's go back to our technical example. If you're looking at the general components that you need in order to run a data center—you have networking, storage, servers, virtualization (this is where an operating system is running within another operating system to make it easier to roll out new services), operating systems, middleware, data, applications—and you decide to be responsible for it all, you're doing it on the premises. You have to buy the servers, install the operating system, set up the virtualization, install the applications, patch, manage, and handle all those pieces and components.

For Infrastructure as a Service, this is where the networking, storage, servers, and virtualization are all provided to you. Essentially, with IAAS, you don't have to worry about hardware. All of the hardware components and pieces are provided by the cloud providers. That means a tech refresh—making sure that you have the latest and greatest CPU and proper storage, among other things—is all handled by a third party. Your company doesn't have to worry about buying any hardware components, but you still have to worry about all the software. You have to install the operating system, worry about the data, the application, etc.

Now with Platform as a Service, the cloud provider is not only going to take care of all of the hardware, the network storage, service, and virtualization, but they're also going to take care of the operating system, the middleware, and all the runtime components (these are the components that are needed in order to have an application work and function correctly). They're going to make sure that that operating system is up to date. When new versions come out, they're going to update it, patch it, make sure that it's fully running and ready to go, and you just have to worry about installing the applications and managing your data and components.

The final level is Software as a Service, and that's where the cloud provider handles everything. They take care of the network, the storage servers, the operating system, the data, the applications—everything. All you need to do, of course, is log in and be able to access and utilize the services.

With the cloud, it's really the authentication and monitoring of that access that becomes critical. It is important as a business leader that when you're considering moving to the cloud, you ask yourself: What level are you looking at, and what level of security is still required and still important?

One of the exercises I always recommend to business leaders is to ask your security people to write down all of the different security responsibilities that are needed for blocking traffic, monitoring traffic, filtering traffic, looking for anomalies, hardening (locking down and a general term used to describe the process of securing a system), incident response, and forensics. Then, make sure it's very clear who is responsible for each issue because probably one of the biggest challenges and concerns that I see when moving to the cloud is a gap. You may think the cloud provider is taking care of monitoring, the cloud provider may think that you're taking care of it, and, in reality, neither of you are. That's why you want your security team to be actively involved with contracts, so that the proper language, SLAs, and metrics are put in place to make sure things are properly handled, controlled, and monitored.

WHERE EXACTLY IS YOUR DATA?

Now that you've learned a little about the cloud and what it is, you're probably asking yourself: Where is your data physically located? If you have an on-premise data center in, say, Ashburn, Virginia, well that's pretty easy—your service is in Ashburn, Virginia, and your data is in Ashburn. But when you use a cloud provider, do you actually know where your data is physically located? Probably not.

A lot of cloud providers have data centers and replication facilities all around the world. This is for disaster recovery purposes—if a server fails in one place, they want a backup elsewhere. For example, if there is a major disaster in California or the entire United States, you would want your servers in a completely different location so they can continue to work.

There was a US government contractor doing work for the State Department, managing and handling a lot of ITAR (International Traffic in Arms Regulations) information. ITAR information is not technically classified, but it has restrictions and cannot leave the United States. This was a relatively small government contractor, and they needed to make sure that data was backed up and replicated. They decided to go to a local cloud provider in order to manage and host the service. They figured it was fine because it was a cloud provider in their own city. What they didn't realize was that the provider had replication facilities in the Philippines, so all of their data was also being replicated to the Philippines.

When the government did an audit, they found out that their ITAR information was now sitting on servers in the Philippines. That violated the export restrictions and the law. This government contractor was making $55 million a year. The fine? $75 million. They went out of business.

I bring this up because cybersecurity isn't just a safety concern, but something to take seriously because it could also put you out of business.

This was a company that was previously growing by 30 percent every year, on their way in three years to grow to $175 million (that's right in the sweet spot of the big government contractors).

They were winning more and more government contracts and in a prime position to be acquired in two to three years, probably for $200 million. Instead, they not only went out of business, but they faced significant fines and penalties that they had to figure out how to pay. It's one of many cases I've seen where ignoring security could be detrimental to your business. It is important to note that in this case, there was no actual data breach, but there was a breach of contract. The company agreed to keep all of the information within the US, and they failed to do so.

A second example concerns a small government entity in Saudi Arabia that had a lot of very sensitive information about that nation, which they did not want any other countries to find out about. Because this was a third-party entity, they hired a local cloud provider that was incorporated and running in Saudi Arabia. Once again, this government agency assumed that because it was a local Saudi Arabian company, all the data would stay within that country. This Saudi Arabian cloud provider needed to make sure that if there was a major disaster or issue in Saudi Arabia that their data was backed up, so they didn't tell their customers or clients that they were duplicating all of their data to a data center in the United States. It didn't take long before the US government figured this out, and they really wanted this classified information. If the servers are in the United States, and the data is in the United States, who has jurisdiction? You guessed it: the US government, which includes the FBI. The FBI was able to go into that data center, physically seize and acquire the servers, and access all of that very, very sensitive information.

Hopefully, you're seeing a theme here: Ask questions. It's critical that you ask questions, put correct language in the contact that specifies where the data is going, and do proper due diligence of all the various cloud providers. Some questions you should ask include:

What is your company's critical data?

At the cloud provider, which servers are the data located on?

Where do those servers physically reside?

What other data resides on those servers?

Do other companies or third parties have access to that data?

Today, many countries do have cloud providers that will keep your data within the country. But, you have to specifically ask, and there's an additional fee because they need to do more work, tracking, and alerting. I don't want you to think that your cloud providers cannot keep data within a region—they absolutely can. But you need to make them fully aware that's what you want, and it'll cost you more.

The big lessons you need to know when it comes to the cloud are what your critical data is and where it's physically located. Remember, just because you're utilizing a company within a country doesn't mean your data stays within that country.

YOUR DATA IN THE CLOUD

Most people start thinking about cybersecurity only after there's a problem, issue, or exposure. As I mentioned before, one of my favorite quotes is, "Smart people know the right answers. Brilliant people ask the right questions." That's why this book isn't a technical manual. My intention is to get you to think differently about security and to do something about it.

When I'm talking about a major breach versus a minor breach, I'm referring to the difference between a breach that puts you out of business and one that is a minor annoyance. The difference is not the system services or who's handling those services—it's the data. You need to become obsessed with asking about your critical data. Whether we're talking about mobile devices in the previous chapter or talking about the cloud in this one, the data is what matters.

What you want to ask is:

What is our critical data?
How is it stored?
Where is it located?
Who has access to it?

And with cloud-based storage:

Where is the data encrypted?
Who has access to the keys?

One of the ways that we protect, secure, and lock down critical data and information is with encryption. I'm not going to get into the mathematical formulas of cryptography within this book, but I'll give you the general breakdown. It's like putting your information in a safe. With encryption, your data is encrypted with a key. If somebody only gets the encrypted data and it's encrypted correctly, it's pretty much useless to them because they're not able to read or access it unless they have access to the key. So the question you need to ask when it comes to encrypted data is where are the keys stored?

In protecting your critical data with encryption or cryptography, it's all about protecting and securing the keys. If somebody can access the keys, they can decrypt your data, and it defeats the whole purpose of encryption. This is probably one of the biggest misunderstandings that businesses have when protecting their critical data. They believe that encryption equals security, so as long as the cloud provider, third party, or their business is encrypting information, they believe that it's protected and secure. What they fail to realize is that encryption doesn't accomplish anything if the keys aren't properly protected, secured, and locked down.

Every company that has had a major breach over the last three years always encrypted that data. The problem wasn't that data encryption wasn't in place—it was that the keys weren't properly protected and secured. Remember, as I mentioned earlier in this book, many companies store the keys with the data. Therefore, you need to make sure that if you're utilizing a cloud provider, you ask them:

How is the data encrypted?
Where are the keys stored?
How are the keys protected and locked down?

The easy answer for you and your business is that the keys must always be stored on a separate server. If you follow these three general principles when it comes to the cloud, then your data is safe, protected, and secure:

1. The key is always stored on a separate server.
2. The key never leaves the key server.
3. The key is never stored with the encrypted data.

There's also another option. With both Infrastructure as a Service or Platform as a Service, one thing you can do is only upload encrypted data to the server. All the crypto keys are stored on your local systems, so even if the data is compromised in the cloud, the cloud provider does not have access to the actual keys that are needed to decrypt the data. The client uses those keys to encrypt the data locally on their systems. At this point, the encrypted data is uploaded to the cloud provider. In this case, the cloud provider only has the encrypted data and doesn't have access to the unencrypted data or the keys.

If an attacker breaks into the cloud provider and they get access to your data, all they have is encrypted data. Nobody can decrypt it, not even the cloud provider, unless they get the keys, which are not on the cloud servers.

Of course, if this is a targeted attack and the adversary can break into the client and steal and acquire the local key, then all bets are off, but that just adds an extra level of protection. So, in addition to asking where your data is being stored, you must also ask:

Who is encrypting the data?

Where are the crypto keys being stored?

What is the overall exposure to you or your business?

THE *F* WORD

I know we mentioned the *F* word—free—earlier in the book when we were discussing mobile devices, but I want to mention it again here because it's also a dangerous word when it comes to the cloud. As a business owner, you want to be so aware of and sensitive to the *F* word. Free is not free. A lot of these free services, whether it's free Google Docs or free Dropbox, are not free. In those cases, your data is the service.

How in the world can they otherwise survive if their entire service is free? If they offer both a paid service and a free service, and there doesn't seem to be a difference, why would anyone ever pay?

With the free service, you agree that they can utilize, analyze, and access your data to perform targeted marketing to you. You probably think you never agreed to that, but you did. Whenever you sign up for those services, it's always in their user license agreement—the one you have to scroll through the tiny print, which no one ever reads, to click the "I accept" or "I agree" button. I've read them and it's in there.

People have tried to sue companies over this, but truthfully, their lawyers are better than your lawyers. They make sure they're fully covered. Now, just so I'm fair to these providers, they do not actually take your data and sell it to a third party. What they are doing is taking your data, reading it, correlating it, and building profile behavior patterns. They're able to create a profile around your interests, health, etc. Then vendors say, "I will pay you X amount per ad if you can go ahead and put my ad in front of these types of people." And the vendors pay them. They're not actually selling your data, but they're still accessing it. The big thing to note here is that if they can access and read your data, other people could potentially get access to it, too.

Free is dangerous. If you want to use free storage for pictures of your kids or grandkids, that's between you and your family members. But from a business perspective, you need to pay money for the services because with the paid services, your data is protected. Whether it's $100, $300, or even $1,000, it's well worth that additional cost.

Trust me. They are watching, monitoring, and seeing what you're doing. Free is not free.

SPECIFIC BUSINESS RISKS

Depending on your industry or the business you're in, you need to be aware of some of the laws that may be applicable to your data. Let's take healthcare for example. There are a lot of rules governing the use of personally identifiable information (PII) and data related to the Health Insurance Portability and Accountability Act. In finance, you have the Financial Monetization Act, among others. Depending on your industry, make sure you understand what type of data you are handling and what regulations are in place, and make sure that your cloud provider is implementing those proper procedures and components.

In the case of healthcare, hospitals and doctors need access to certain data. They try to make it easy, but there are often risks. If you work in healthcare, it's important to ask when evaluating access to data:

What is the value and benefit?
What is the exposure point?
Can we live with that risk or exposure ... and if we can't and we need the functionality ... ?
How do we mitigate that risk to an acceptable level?

And then, of course, there's the confidentiality, integrity, and availability to consider. You want to make sure you're protecting the data from inadvertent disclosure, which is defined as confidentiality; alteration, which is defined as integrity; or denial of access, which is defined as availability.

CHAPTER 5 REVIEW

When your data is moved to the cloud, it's controlled by a third party. Cloud services, though, may actually offer better security because they're dedicated to monitoring and tracking your data very closely since it's their primary business to safeguard your information.

Choosing to store your data in the cloud doesn't necessarily mean that you can reduce your security staff; you need to decide what level of cloud service works best for your company. For example, if you choose Infrastructure as a Service, you're probably still going to need a large security staff. If, instead, you move to

Platform as a Service or Software as a Service, your own in-house staff can be more hands-off, and you will need less direct staff to manage your data storage.

Ultimately, you want to assure that whatever cloud provider you put in place, they are still hardening, locking down, patching, and securing your systems. You need to know that they're constantly monitoring your data, looking for signs of an attack, and there is proper authentication in place.

The bottom line is that, before you sign any contract with a cloud provider, you put proper security SLAs in place, which make very clear who's responsible for all aspects of the security of your data.

If you ask the right questions and have the right mindset, the cloud can actually be a very nice addition to your business and reduce security costs. But you need to understand the risks and the exposures, and put measures in place to minimize them.

SIX

THEY'RE IN YOUR BUSINESS

Whether you like it or not, if you run a business today and you have any information stored in digital form, not only are you a target, but the probability that you have been compromised already is as close to 100 percent as you can get.

Remember the two fundamental facts: You are a target, and cybersecurity is your responsibility. I could write volumes of books showing you that it doesn't matter who you are, what you do, how big, how small, whatever type of business you're in—you're going to be a target. Everyone has critical information and someone is going to target you.

Even if you have third-party services in place, cybersecurity is ultimately your responsibility. You can transfer functionality, you can even transfer security, but you cannot transfer liability. Let's go deeper into what you need to be aware of when it comes to your business and what you can do to protect it from cyberattacks.

WHY YOU ARE A TARGET

You may be a target for many reasons. First, it could be based simply on the name of your business. A friend of mine, who had a fairly small business, got attacked disproportionally. The name of his company was Apple Technologies, but it had nothing to do with the Apple corporation. There were no violations in the name because it was in a different sector and a local business, but guess what? Attackers don't know that. If you have somebody in Russia, Venezuela, or some other part of the world, and they want to target a large tech company like Apple, they're going to start searching and attack what they find.

Guess which bank in the United States is targeted more than any other bank? Yes, it's Bank of America, and the reason is simple: Most other countries have banks associated with the country. There is the Bank of Mexico, the Bank of Singapore, the Bank of Brazil, etc. These are run by the actual countries. Many foreigner attackers don't realize that is not how the United States works, and our equivalent is basically our federal reserve. But they think Bank of America is "the Bank of America" and for those reasons, it's a big target. It's why your business name alone can paint a target on your back.

The second reason you are a target could very often be the country you're located in. There are some people who do not like the United States, so just by being a company incorporated in the United States, you could be targeted. The location of your business is public information, so it's easy if someone wants to target you just for your country, state, city, or certain area.

Then you could be a victim based on the type of business that you're in. I know oil and gas is a business not many people like because of a belief that it's not environmentally friendly, so it gets hit a lot. Does your business have anyone who could potentially oppose it or dislike it? Be creative here because whether you like it or not, one of the things I found in cyberspace is that everybody has enemies, and the more that you're aware of them, the better off you'll be.

The next reason you could be targeted is because of who your customers are. I know that a lot of my friends who have small businesses are often in the crosshairs, based on the government agencies they deal with that some people do not like. I have a client that does work for the National Rifle Association, and there are people who are against guns or the association, so they target their company just for that reason.

And then finally—and this is probably the most important one overall—you are often targeted based on the information or data that you have. This can be broken down into several categories. The major one is if you possess intellectual property. Every business has some information that gives them a competitive advantage, even if you don't realize it.

I consulted for a CEO of a $45 million company who didn't really get the importance of cybersecurity. He said to me in a meeting with his staff: "Eric, there's absolutely nothing special about what we do. If somebody got into our computers, took all of our information, and put it all online, it wouldn't really matter."

I called his bluff. "Cool, let's do it."

The room got quiet.

"Let's go and put everything online," I said. "I could save you a lot of money that way if that's what you're worried about. Forget passwords, authentication, or firewalls. We can cut all that stuff out."

And after a couple of minutes, he sort of got it and said, "Okay, well, maybe we need a little protection."

You have more sensitive information than you realize. The information of your employees and customers is also incredibly valuable. It's easier for an attacker to steal two hundred personally identifiable records from you than to try and get them one by one. Remember, today's attackers are not going after one company, trying to steal fifty million records. Their ideal attack is to target a million companies and steal fifty records from each. It's actually much easier and beneficial in many cases—and almost never reported, much to the attacker's advantage.

Another point, which few people think about: Who is in your supply chain? I used to work at Lockheed Martin, a large company with billions of dollars of contracts and intellectual property; they put plenty of resources into security. It would be very difficult for a hacker to break in and try to steal information about their military contracts. It wouldn't be impossible, but extremely difficult—and attackers know that.

THEY'RE IN YOUR BUSINESS

What about a small company that has $8 million in revenue and fifty employees? How much money and time do you think they spend on cyber protection? Very little.

In this particular case, I worked with a company that size in which it was actually the owner's son who spent four hours a week "doing cybersecurity for the company." They didn't focus a lot on cybersecurity because they thought they didn't have the revenue to justify it.

This company specialized in and had a patent on a very unique chip. This chip was sold to another vendor that went on a computer cart, that went to another vendor that went into a computer, that went to another vendor for the navigation, that went into another vendor. Five vendors later, it wound up at Lockheed Martin and was put into one of their airplanes.

What do you think the attacker did? They went after that small company. They broke into that company, and then from there, they were able to ultimately work their way upstream to be able to break into that bigger company.

You can see how supply-chain verification and understanding is very, very important. Whether you're on the giving end if you're a small company or the receiving end if you're a large company, understanding your supply-chain cycle could be vital.

CYBERSECURITY AS PART OF YOUR BUSINESS

A lot of executives that have big cybersecurity teams think they're set. They figure someone knowledgeable in this area is sure to take care of all of it. Why in the world would they worry about cybersecurity? As we discussed, we can give you the safest car on the planet, but if you're not a safe driver, it's not going to protect you.

Also as mentioned earlier, as I'm sitting here writing this chapter, we're in the middle of the COVID-19 pandemic, and I hope and pray by the time this book comes out that we will have this fixed and things will go back to normal. But right now, many companies are not allowing people into the office. They're switching to a remote workforce, which emphasizes the importance of cybersecurity, because a lot of these people who were traditionally in an office are now at home. Executives just like you, reading this book, used to be behind six or seven different filtering security devices to protect them while they were in their office. But now they're home, using a home computer with direct access to servers, and all that security is gone or reduced. This really emphasizes that cybersecurity is your responsibility.

It is important to remember that it's not the operating system's or machine's responsibility to protect you. Microsoft, Cisco, Amazon, or other cloud-based hosting companies . . . they can give you a really secure system. But the most secure system is one that has no functionality because when you add functionality, you're going to reduce security. Even the Microsofts and Apples of the world know that cybersecurity is your responsibility. As noted, they can give you all the tools and methods and built-in security, but in order not to impact functionality, they don't turn it on by default. It's your responsibility to understand and recognize that you need to go in and turn on the security.

A perfect example of this is that many people are criticizing Zoom because unauthorized people are joining calls and making inappropriate comments. What they fail to realize is that there are security features built in that will prevent this from happening, but the person running the meeting has to go in and turn on those features.

THE FIGHT YOU'RE UP AGAINST

I know it to be a fact that attackers from many countries, including China, Russia, and North Korea, are in your systems. We know who they are, and in some cases we know their business, their name, address, IP address—everything.

Here's the problem: In those countries, hacking outside the country is not necessarily illegal, and we don't have extradition treaties with those nations. So we know that it is happening, but unfortunately, because of the laws, we can't do anything about it.

If you're frustrated that we can't just go in and arrest them, you're not alone. But why is that? It's because the world was created on physical boundaries, physical countries. Before the internet and cyberspace, if you were going to commit a crime in a country, you had to physically go to that country. If you stepped off the airplane onto a country's soil, you knew you had to abide by their laws.

The problem with the internet and cyberspace is that national boundaries are gone. I could go anywhere in the world on the web and sometimes not even know where I am. The next time you surf the web and are on a website, ask yourself: Where am I and what laws am I following? I will tell you that in most cases, you don't know. Your connection might be going through Canada, Germany, South America, and you might actually be visiting a website in Russia. That means that criminals can commit crimes anywhere in the world, and it's very hard to stop them or do anything about them because of those laws.

We can block their IP addresses and even work with some of the cloud providers to try to shut them down. But these cyber-criminals would just pop back up again. It's like whack-a-mole—boom boom! And it's an unfair whack-a-mole where the moles pop up a lot quicker than you can whack them. It's almost guaranteed you can't win.

The UN is trying to work on this, but honestly, just knowing how long it takes countries to work together and all of the technicalities involved, we're probably talking eight to ten years before there's an international cyber police that could actually shut down, arrest, and minimize these issues and crimes. And because it's a problem that is not going to be solved anytime soon, it just goes back to what I've been saying: Cybersecurity is your responsibility.

WHAT HACKERS WANT

Let's take a look at why hackers would want to attack you and what's at risk. Almost every company at some level has competitors. Even if you're an executive with the government, you still have a budget, and you're competing with other government agencies for money and resources. A lot of what I am going to talk about here is focused on commercial business, but all of it applies to government, nonprofits, or any other entity.

I want you to take a close look at your business. What business are you really in? How do you make money? How do you get new customers? What makes you unique? What are your profit margins? Because guess what? That's all intellectual property that competitors, foreign countries, or cybercriminals want.

We are living in a world where cybercrime is overtaking traditional crime. I've asked executives, "How many of you would actually hire somebody to physically break into your competitor's office, potentially eliminate threats of people or guards, and physically steal customers' printed records or intellectual property?" Maybe 1 or 2 percent of the people say that they would (which is scary, but a different discussion).

Then, I ask the same group, "Okay, how many of you would think of potentially hiring somebody to maybe go online and see if you can find information, even if you have to possibly break through a few firewalls, guess a few passwords to gain access to it?"

And the excuses start pouring in: "But Eric, nobody is going to get hurt. You're not going to physically have altercations or anything like that. You're just sort of accessing information. And what if we don't technically break in, but just take advantage of flaws like an attacker would? If their system is weak, is it so bad to just exploit it? What about that?"

And the scary part is, after all of the excuses, 40 percent of them would consider it.

There are some serious moral and ethical lines blurred very quickly, but competitors are viewing this as a viable alternative. As we're living in a world deeply impacted by COVID-19 right now, where the economy is tanking and businesses are struggling and going under, and nobody's buying and everyone's looking for new customers . . . Well, I've been in this game for thirty years and I am confident that not only are cyberattacks from criminals and countries going to increase, but cyberattacks from competitors are also going to start becoming more commonplace, and you're going to see even more and more people targeting and going after you.

I want you to spend a few minutes asking:

What are the things that are really important to your business?

What is the data, the information that gives you a unique advantage?

What are your customer lists?

Where are your employees?

The more you can identify the areas of your business that are important, the better you can build a strategy to put better measures of protection in place.

There's a major gap when it comes to security teams solely being responsible for safeguarding your business: Most security people who are making recommendations to protect your company don't understand the business.

When I teach classes to security leaders and chief information security officers (CISO), I always ask them, "How many of
you know how to read a balance sheet? How many of you know
how to read a profit-and-loss statement?" It's maybe 20 percent.

And I say, "How many of you, before you make security decisions, actually look at the company's financials, look at how they
work, and how they operate?" It's less than 5 percent.

I sit there and shake my head. How in the world can you
determine how much money to spend on security if you don't
know how much the assets are worth?

Although it depends on many factors, cybersecurity is usually 2 to 3 percent of overall revenue.

Let's say we have Company X, which is wondering if $1 mil-
lion is an appropriate security budget. Well, it depends. If this
company only makes $800,000 in revenue a year and wants to
spend $1 million, that's not going to work. In that case, maybe
your security budget is $30,000. On the other hand, if I had sug-
gested a million-dollar budget at Lockheed Martin, I would have
been laughed out of the office for recommending something
so paltry. That number needs to be a lot bigger based on all the
assets and all the information.

What is the data or information that accounts for most of
your revenue and most of your profitability?

As a business owner and executive, you should know that
information. The communication channel between executives
and the cybersecurity team is broken. This is an area where you
can actually help your business regardless of whether it's small,

medium, or large, and work with your security team to help them identify what the critical assets really are.

HOW HACKERS CAN HARM YOUR BUSINESS

I knew a dentist who had three offices and was doing very well. He was making a lot of money and planning on opening two more offices every year for the next three or four years. He was a great dentist, and everyone in the community knew and loved him.

I used to joke with him about coming to me when he was ready to secure his customers' records, but he assured me his system was good. Then I remember running into him at a football game, and he was off to the side, acting very strangely.

I approached him and he said, "I guess you didn't hear. We were hit by a cyberattack and they stole forty thousand of our customer records."

He pointed to everyone in the stand.

"All of our friends, relatives, and neighbors . . . even though many of them understood and recognized that it wasn't my fault, they're no longer customers."

Within a forty-eight-hour period, they went from being one of the top businesses in the county with unbelievable growth and an amazing future to literally being out of business with most of the community frustrated and angry at them.

He said, "I wish I had listened to you, because I didn't think that we would be a target, and I thought the third parties we were outsourcing our IT to were taking care of security and they weren't."

I knew a medium-sized law firm that was doing about $400 million a year. They captured the market in a niche area and were growing. But once again, their whole business was based on client confidentiality of information and records. They had four very big clients that covered 90 percent of their business. Each one was worth about $80 to $90 million.

They were spending $20 million a year on security, but nobody ever checked, verified, and validated, and it wasn't very secure, and they got broken into. Those four clients had all of their critical information publicized. Not only was it embarrassing for them, but there were lawsuits. This happened two and a half years ago, and today they are fortunately still in business, but they have a $12 million business instead of a half-billion-dollar business. Once again, they didn't think they would be a target, and they didn't think that cybersecurity was their responsibility.

And then we have the example of a large business. This was an international chemical processing company that had facilities all around the world: in Europe, the United States, Russia, and China. Once again, they thought they had the basic protection in place and that was enough. It boggles my mind that this was a huge company and they only had three people doing security, and these people really were in IT. They weren't properly trained. I'm not being critical here, just stating the facts. They were paying almost $40 million a year to third-party vendors, but nobody was managing or tracking those vendors and their work.

They started noticing that their competitors, especially in China, were beginning to produce similar products and technologies. When they looked closer, they realized the only way those competitors could be doing it at those margins is if they had their technology.

They brought me in and we did a threat-hunting exercise. Generally, threat hunting is a paradigm change: You assume you are compromised, and you actively look for the adversary. We found that multiple executives, including the laptop of the CEO, had been compromised for over seven months and nobody knew about it. Not only that, but many of their critical servers were also being monitored and tracked.

It's sad to say this, but they actually were better than the industry average. They were compromised for about a year and a half. Fortunately, they called us in and we were able to find those compromises. Most companies are compromised for three or four years without even realizing it.

I remember doing the final briefing with the executives, and we taught them to ask better questions and act like every day someone is going to attack them. Unfortunately, this is the norm: Many companies do not focus on cybersecurity until after they have a breach.

WHERE ARE YOUR EXPOSURE POINTS?

You think whether or not you'll get attacked is a gamble, but it's not. Well, small companies are a little less likely to get attacked. Medium, probably will. Large? Absolutely. The bottom line is that all companies are going to be attacked.

I remember at a super-large Fortune Ten company, the executives weren't getting it.

I said, "Let me ask you a question: What if I told you there's a room that has thirty desks in it, and there are thirty full-time people. They come to work at eight o'clock every morning and work until six or seven every night. They get benefits, salaries, and their sole job for the next three years is to break into your company. That's their only job. There are thirty dedicated people who are doing nothing but spending every single day targeting, looking at, and going after your company. That's the threat you're up against."

They got it.

Just to show you that these attackers are in your networks, and just how advanced they are, let me give you another example. This was what we call an APT case, an Advanced Persistent Threat. This adversary is very savvy, persistent, patient, and they're going to take whatever time is necessary to target and break into your organization.

At a medium-sized $800 million manufacturing company, a vice president in charge of research and development was targeted by a foreign adversary. As far as we could tell, the company was being targeted for seventeen months before the attacker took any aggressive action to break in.

Think about that: seventeen months! That's pretty persistent! That's prodigious waiting and waiting and waiting for the opportunity. They found and targeted this VP because he's very present on the internet for social media and conferences, and so they learned everything about him. They knew where he lived, where he grew up, found out about his wife, his kids, his family—everything possible. They monitored him for seventeen months, just waiting for an opportunity to break in, until they found it.

There was an accident at his son's high school. This adversary had nothing to do with the accident, but within four hours of the accident being broadcast all over the news, this VP received an email from the superintendent of the school district.

It said: "Dear concerned parent. As you are probably aware, there was an accident at Eagle Ridge High School. Our utmost and primary concern is the safety of all children involved. We are in a lockdown state, so please do not come down to the high school. Attached is emergency contact information where you can find out further details. We appreciate your continued support. Sincerely, the superintendent of the school district."

This email was flawless. No spelling or grammatical errors. It looked polished and legitimate. Everything was perfect. When you opened the attachment, it was on the school letterhead and had all the correct, valid information.

The VP opened it up and he didn't think much of it again. It wasn't until eight months later that his IT department noticed unusual activity on the network and some strange things happening. They called us up and brought us in to investigate, and we determined that not only were they compromised, but we traced back the root cause of the attack to be that VP's computer.

It turns out that email wasn't from the superintendent of the school district, but from the foreign adversary. The VP was floored.

He said, "Eric, what kind of adversary are we dealing with? My son could have died. Kids lost their lives on that day, and these people are going to exploit that to break into my company?"

And I said, "Unfortunately, yes. That's the adversary you're dealing with. Not only are they in your network, but they are targeting you and coming after you, and doing whatever it takes."

Knowing all this, let's look at the areas you specifically need to be aware of and responsible for. Remember that when we talk about cybersecurity, there are three main things we want to look at:

The risk: Cybersecurity is about understanding, managing, and mitigating the risk of our critical data being disclosed, altered, or denied access to.

Critical data: Understanding what and where that data is.

CIA: Confidentiality, Integrity, Availability. Confidentiality is preventing, detecting, and deterring the unauthorized disclosure of information. Integrity is preventing, detecting, and deterring the unauthorized alteration of information, and availability is preventing, detecting, and deterring the unauthorized denial of access.

Some of the things I want to break down we've briefly touched on in earlier chapters, but they're worth repeating because I'm going to cover them from a different perspective.

What I find interesting is that most executives are actually very good at understanding and managing every other business risk, except for cyber. They don't realize or see or think about how somebody could break in and steal their customer records or critical intellectual property. One of the things that I do when I work with clients, which I train all my CISOs on, is that when you present findings to an executive team, you should create a chart that has columns that state:

What is the risk?

What is the likelihood of it occurring?

What is the cost if it occurs, and what is the cost to fix it?

I urge you to ask your security team for information in this format to help you better understand the realities of your cybersecurity risks. You want to know your top risks:

What is the percentage chance that this can occur?

How much money are you going to lose if it occurs?

How much money do you need to spend in order to fix or stop or reduce this from happening or occurring?

That's the information that you want. That's the data you want to ask for.

Remember, whenever you're making business decisions, there are always going to be security risks: 100 percent security means no functionality. So any time you're expanding functionality—anytime somebody is asking for a new application, a new website, a new service, a new component—there will be security risks. You need to understand: What is the value and benefit of the functionality you're getting? The exposures and risks it creates? Can you live with those risks? And if you can't live with those risks, what can you do to reduce those risks to a more acceptable level? That's how you have to think now.

Then the second area you need to identify is your critical data. Go back to what we did earlier:

What business are you in?
What is your competitive advantage?
What are your unique offerings?

Many times companies don't know their critical data, and therefore they're not protecting it until after it's compromised.

We worked with a large hotel chain client that bought a smaller hotel chain, and they migrated all of their data over to the new system. However, nobody ever recognized that the databases from the company they acquired still had live customer data in it, and they'd never decommissioned the system correctly. Attackers didn't break into the primary hotel database because that was secure but instead broke into the older database that still had critical records that nobody identified as being important data.

One of the best ways to secure your data is to reduce your footprint. If there is data that you don't need, that's not required, you need to remove or decommission it. In security, we call this "reducing the attack surface."

If your company hasn't done it—and very few have—I recommend you do a data discovery exercise. Ask your IT or security team if they know where all of your data is. When I ask companies if they know what their critical data is, and where it is located, they can almost always tell me which data is critical and on which servers it is located. However, I've found that it's almost always also located on other servers they didn't know about.

If you believe your data is only on a few servers, what are you going to secure? Those servers. If it's on seven other servers that you're not aware of, that you're not protecting or securing, which do you think the attacker is going to go after? Obviously the ones you're not guarding. So most companies can identify their critical data and where it's located, but they are not aware that it also resides in other locations, and therefore they are not protecting it the way that they should.

Another way to secure your data is to prioritize which of it is critical. Determine what is most important, making sure that it's not disclosed (confidentiality), altered (integrity), and you can't be denied access to it (availability). Start prioritizing that. As an executive, you should know your critical data, the risks, and which area of the CIA triad is most important for you to protect and secure.

Hopefully by now, reading this book, you have some better questions to ask your staff. To round that out, what about:

If you were compromised, how would you know?
How much confidence do you have that if you were compromised you'd be able to detect it?

If your staff tells you that they're 100 percent confident that if you were compromised you'd be able to detect an attack within two weeks, that's simply not true. The best I can say for the *best* companies out there is that if they were compromised, they would be able to detect it with 85 percent certainty within two months. *Most* companies, if they were compromised, would be able to detect it within about two or three years, and only then would they catch about 50 or 60 percent of it. The main reason we miss so much is because there is usually no visible sign of an attack. It's the same reason why many people do not detect an illness in their body until it is too late. In some cases when people are really sick, there is no visible sign. They feel healthy and perform their normal activities. It's not until they start having symptoms that they realize they need to go to the doctor. In some cases, the doctor will tell the patient that if he or she had come in sooner, they could have found it earlier and treated it. This is the reason why routine physicals are so important, and why we regularly go to the doctor even if we are healthy—to get checked.

The same applies to cybersecurity. In the late 1990s, when cyberattacks started to become more prevalent, they were visible and easy to detect. Like with the ILOVEYOU virus that we mentioned early in this book, you came into work and your inbox was filled with ILOVEYOU messages, and it was very obvious that you were compromised. When I worked at the CIA, hackers defaced the website to make it say "The Central Stupidity Agency." It was very clear that we were attacked. But today, the attacks are much more stealthy so people don't notice them. From a company's perspective, everything looks fine and normal, and since there are no signs of an attack, companies assume that everything is fine. If someone broke into your company and slowly copied sensitive data out of your organization, how would you notice? Unfortunately, unless they started overloading the servers (as mentioned earlier, the most common way hacking is detected), there would be no visible sign of attack and no reason to think that you are compromised.

This is why I emphasize threat hunting with my clients. With threat hunting, you assume you are compromised, and you actively hunt and look for the adversary. Are you looking in the right spots and is your security aligned with how the adversary works?

One of the questions executives always ask me: What are my big clients really using to catch these high-end attacks? Almost all of them rely on the exact same thing—the FBI. Almost all of the big compromises that we've been involved with were notified by the FBI. It's usually because the FBI was doing work on another project and accidentally discovered the company's data where it shouldn't be. They call the company, and the company calls us. I often joke that I wish I could pay the FBI agents a commission because they bring in more business than my salespeople do.

I love the FBI and I think it's great that they do that, but their job is not to detect when your network becomes compromised. That's your company's job. So just keep asking yourself, how confident are you that you will be able to detect a compromise? Are you looking in the right spot?

One of my favorite examples of this is from more than ten years ago when I started traveling and doing a lot of teaching and giving keynotes. I went to Singapore for a week, Australia for a week, and then ended up in Hawaii for the final week. I got to Hawaii two days early so I had a day off to relax, something I hardly ever do.

I was traveling with a CEO I knew, who later turned into one of my mentors, who said, "Eric, let's go whale watching. It's relaxing to me. You can sit on the back of the boat, we can talk, catch up, get some sun. It'll be great."

It was beautiful. Only we hadn't seen any whales. I kept looking out on the left side of the boat where we were sitting, and nothing. As soon as I mentioned it, almost like on cue, he started to get excited.

"Woo-hoo! Look, look!" He literally put his hand on my head and turned me to the right side of the boat where there were the most amazing whales I've ever seen.

If I had been on that boat without my friend John, and I had been looking at the left side of the boat only and didn't see any whales, would it be safe for me to say there aren't any whales in the ocean? No. We're pretty much sure there are whales in the ocean; we're just not always looking in the right spot. It wasn't until John changed my focus and my point of view that I was able to notice the most beautiful whales I've ever seen.

That's how many companies approach security. They're looking at some high-level logs of firewalls and some general devices and don't see any attackers, so they assume they're not there. If you think your network is attack free because you're not seeing them, that would be as naive as me saying the ocean doesn't contain whales because I didn't see any on that particular trip.

If your staff is telling you that there are no attacks on your network, have them do something different. If your staff is not finding breaches, tell them they're not looking in the right spot. If you say, "I'm 100 percent confident we're compromised. Find it and don't stop until you do," it's a completely different perspective than approaching it like you're probably not compromised, and then maybe you have a shot at actually finding the compromise. The right approach is that you're almost guaranteed to be compromised. Keep looking, keep looking, keep looking until you see some whales.

To find your exposure points, I want you to start thinking like the adversary. Let's say that you woke up and you're now a cybercriminal, or let's say that you hate your current company, and you turned evil, and you now work for a competitor and want to steal your old company's data. How do you do it?

One of the reasons I'm so good at cybersecurity is because I spent eight years as a professional hacker. I think and act like the adversary. It sometimes freaks out my family and my friends that I think that way, because I'm a little paranoid and crazy, and I'll admit all that. But I always think like the offense. To be good at defense, you need to understand how the offense works.

The more you can think like the adversary, the better prepared you'll be. From a company perspective ask:

What data do you currently have that an adversary would want?

What is the data that if a third party, a competitor, or a cybercriminal got ahold of, could not only be very, very valuable to them, but also be very, very impactful to your company?

From the attacker's perspective and thinking like them, you would ask:

How would you break in?

How would you steal it?

Many of the attacks are a lot simpler than you realize. With a lot of our clients, one of the ways that competitors steal information is to target you at an airport and steal your laptop. Very low tech, very simple, but very effective. Another one that we've seen is phishing attacks, which we described earlier in the book. Once again, simple but very, very effective.

You don't have to be super technical, but let me give you some of the answers to help you out. Your critical data is typically broken down into the following categories:

Personally identifiable data: These are usually records about your employees and your customers.

Intellectual property: You're going to have certain ways that you do things and certain ways that you deliver that give you a competitive advantage.

Methods of delivery: The way that you deliver and service your clients also represents significant intellectual property.

Then how would you break in?

Over the years, I have worked on hundreds upon hundreds of breaches. The main exposure points are systems accessible from the internet that aren't properly protected, secured, and do not contain critical data. You want to start looking at those unpatched, outdated systems that have known vulnerabilities, known exposures, and known points of compromise. Then find the unprotected data, critical data, and critical information that's not properly protected or secured. Finally, find those emails that are malicious but look legitimate, which have attachments or links in them.

WHAT YOU CAN DO TO MINIMIZE BEING COMPROMISED

Make sure your staff accurately knows all the servers and systems that are on your network, how they're configured, and that the staff assesses the overall risk and exposure. The foundation for security is asset inventory and configuration management. If you don't know what servers are on your network and how they're configured, you're not going to win.

You want to be careful of what I call "the curse of 90 percent." If you look at all the big companies that have had breaches, do you think they had no asset inventory and no configuration management and no clue? No, of course not. But here's the problem: They knew 90 percent of their assets. They had 90 percent of them secure, and they knew where 90 percent of their data was. You'd think 90 percent, that's pretty good—but not in cybersecurity. That 90 percent means a 10 percent loss, and that's where the compromise occurs.

I know I told you you'll never be able to have 100 percent security if you have any functionality, but you can have 100 percent asset inventory and 100 percent configuration management. You need to make sure you are actually doing 100 percent in the areas that really count, because anything less is an exposure to the adversary.

I want to finish up with a quick exercise I do with all of my clients. Start by asking your team: What are the biggest threats to our organization? Have them tell you the top five or six threats that could cause the most harm to your company. Then ask: What are the vulnerabilities that would allow those threats to manifest themselves? And finally: What is the critical data that would allow those threats to exploit those vulnerabilities and exploit that critical data? Once you start understanding what those vulnerabilities and exposures are, you want to make sure that they're getting fixed and remediated on a regular basis.

Once again, I don't expect you to be a technical person, but if there's something that could put you out of business, you need to identify it and be aware of it. Some of the metrics that you should have your staff track are:

- the number of visible systems
- the number of outdated systems
- the location of your data
- phishing stats (how often your team clicks on links or opens unauthorized attachments)
- overall impact to your system and organization if an attack occurred

WHAT YOU CAN DO IF YOU ARE COMPROMISED

You might now be thinking: But what can I do if I am compromised? First and foremost, if you believe that you have been compromised, you want to contact cybersecurity professionals who can help walk you through the process. Don't try to deal with it yourself because often simple things like turning off a computer or disconnecting from a network can destroy evidence and cause more damage. Similar to if you are sick, you can try self-diagnosis and take over-the-counter medicine, and if it is a minor cold, that might work, but for any serious illness, you need to go to the doctor. A cyber breach is no different. If it is minor, you might be able to deal with it yourself by resetting some passwords or reinstalling your operating system, but if it is any type of major breach, you need to get help from a professional.

It is also important to note that depending on the type of crime, there might be mandatory reporting requirements that must be done within a certain time period, so it also might be important for you to consult a cyber attorney who can provide accurate advice on whether law enforcement needs to be notified and/or you have to notify your clients, depending on the situation.

CHAPTER 6 REVIEW

You could be a target simply because of your name, location, type of business you're in, the customers you serve, the information or data that you have, or who is in your supply chain. Help your security team identify critical assets so they can help you protect those assets. When it comes to your business, in addition to accepting that cybersecurity is your responsibility, ask better questions, and act like every day someone is going to attack you. Your company will be targeted because it's somebody's sole job to get in and steal your data.

When identifying your exposure points, ask: What is your risk, and what is your critical data? One of the best ways to secure your data is to reduce your footprint. The other is to go back and prioritize your critical data. You will be attacked, which is why detection is so important. Are you looking in the right spots and is your security aligned with how the adversary works? If you think you've been compromised, be sure to contact a cybersecurity professional.

SEVEN

NATIONAL INFRASTRUCTURE ATTACK

Our nation is currently at war. Whether we realize it or not, we're in the middle of World War III. The reason why many people don't recognize it is because it's a different type of world war. In this war, every single country is involved, and every single country is both being attacked and attacking other countries. It doesn't involve bombs, weapons, bullets, aircraft, or navy ships, but instead involves packets, the internet, and computers, which can be far more impactful. We are at cyber war. Whether we like it or not, countries are actively targeting and attacking each other to disrupt both intellectual property and national infrastructures.

What I find particularly interesting is there are no allies. In traditional wars, we have had allies. If the UK was attacked, the United States would send forces, troops, and reinforcement. But in this cyber war, it's every country for itself. We track and monitor all the attack vectors coming in and out of different countries, and the United States is attacking Canada and the UK and Australia just like the UK, Canada, and Australia are attacking us. It's a different type of war because every country is actively participating in its own way, shape, or form. It is important to note that while the attacks are coming from these countries, we do not necessarily know who the actual hackers are or if they belong to any sort of organization. While in some cases, like North Korea, most if not all of the attacks are coming from the government or sponsored by the government, in other cases like China, it is mixed. Attacks coming from Russia are typically done by criminal elements with some financial gain or benefit as the motive. Overall, both government entities and private criminal elements are launching cyberattacks on a regular basis.

The scariest part is that we don't really know how bad it is. We just know it's bad. We know for a fact that many countries like Russia and China are in our infrastructures. They have compromised computer systems within our country and established footholds. But I will also tell you that we are actively attacking them and also have a foothold within their infrastructures and their countries. So we know that it's bad, but the problem is we don't know what we don't know.

We do know that some attacks are occurring, so that's the "known known." We also know for a fact that there are some attacks that we're missing, that we can't possibly be catching every attack out there, so that's the "known unknown." What really scares me is what I call the "unknown unknown."

I often use the iceberg example. Above the water, the piece of the iceberg we see is the "known known." We know that there's some part of the iceberg under the water that we don't see, which is the "known unknown." The "unknown unknown," however, is the size below the surface. With icebergs, sometimes they're much larger below the surface, and sometimes they're not. We don't really know what we are missing here, and that's what worries me the most.

I know you might think I'm exaggerating, but consider it for a second. When I say "information warfare," what countries come to mind? China, Russia, North Korea. Just so nobody gets mad at me, we can put the United States on that list. But the point of it is this: No matter how many people you ask, China is always in the top three. The fact that the Chinese are actively trying to target and steal our intellectual property is not a secret or surprise. As noted earlier, take your electronics—camera, TV, monitor, speaker, computer, hard drive, cell phone—and flip them over and on the back side you should see three words that bring warmth and comfort to your soul: "Made in China."

Just let that sink in for a moment. If we are concerned that the Chinese are actively targeting and breaking into our systems, yet all of our technology and all of our electronics are made in China, what makes us think for a second that they're not embedding technology into those devices? How would we know? Most companies are not opening up laptops, or verifying, validating, and checking those laptops. And it's not uncommon for embedded components that listen, monitor, or take information to have been built in and not activated for five, six, or seven years before anyone notices or takes action. A lot of this stuff is passive. They just embed it. If I was going to launch an attack against the world, or let's just say against the United States, I would start today with all chips, all computers, and everything I'm manufacturing, embedding back doors and components that I don't plan to use for ten years because no one is going to suspect that, and by the time they catch it, it will be too little too late.

There have already been multiple cases where electronics coming from China have had embedded components in them. Just a few years ago, there were twelve thousand commercial laptops sold that turned out to have monitoring chips built into the hardware that were just monitoring and gathering information.

A few Decembers ago, during the holiday season, one of the big-ticket items was digital picture frames. You could buy one, load pictures, and then give it to grandma or grandpa. Well, it turned out that over thirty thousand of those digital picture frames, shrink-wrapped in boxes, were embedded with malware. Talk about a gift that keeps on giving. You plug it into your computer to upload the pictures and you get infected. You then give it to grandma and grandpa, and they plug it into their computer to update the pictures and *they* get infected. The malware was embedded during the manufacturing of the actual devices. It is hard to tell whether this was intentionally done by the manufacturer, sponsored by a foreign government, or just a cyberattacker going after the supply chain. Regardless of which one it was, it is happening. Supply chain attacks are becoming more popular because instead of trying to break into thousands of individual computers, if an attacker can compromise the supply chain in one spot, everyone who buys the device will be impacted.

Probably the scariest of all was a case with five thousand Cisco routers sold in the United States. They looked like Cisco routers, came in Cisco boxes, had the logos, everything looked perfect, except there was one issue: They weren't made or manufactured by Cisco. They were actually manufactured by a competitor in China.

Now, if all they were doing was basically stealing the technology and selling it under their name and label and making money on it, yes, that would be problematic. And if I were Cisco, I'd be very, very upset. But if the routers worked exactly the same way, it wouldn't necessarily have a huge impact on those individual companies. However, it wasn't until a couple of years later that an engineer was actually working inside one of the routers and noticed an additional chipset. After the investigation, it turned out that these five thousand routers had an extra chipset that was actually monitoring and watching incoming and outgoing traffic. These routers were in government facilities, military bases, commercial companies, Fortune 500 companies—all of them had infected routers that were monitoring and gathering traffic. Five years ago, a notice was put out to have people check for these routers.

Here's the scary part: There were 5,000 routers that we know about, and from the last count, only 3,200 were actually turned in. That means there are still 1,800 of these infected routers installed at companies and organizations because nobody checked.

This idea that we are under attack from a national infrastructure perspective is not a Tom Clancy fiction novel. This is reality. And the problem is we only find out about these after we look very, very hard. Therefore, it's incredibly likely that it's a lot worse than we realize.

When I say that the Russians are in our infrastructure and we're in theirs, this scares a lot of people. They start thinking about how the Russians could take down our electricity, utilities, or cause massive havoc. Technically yes, but to me, the war we're in is a cyber cold war, as we described in Chapter 1.

Just to show you that critical infrastructure is targeted and at risk, I want to go back to the example that happened in Iran known as Stuxnet. To refresh your memory, Stuxnet was a cyberattack in which one of the Programmable Logic Controllers (PLCs) that runs a nuclear reactor within Iran was actually hacked and essentially caused it to overspin and melt down the reactor. It's a pretty serious attack when you're going after nuclear reactors.

What was interesting is that PLCs in nuclear reactors are not connected, or not supposed to be connected, to the internet. Well, it turns out that the specific PLC they were using actually had the malicious code embedded in it at manufacturing time, and this particular PLC was one of thirty devices that were sold. They only checked the thirty that were part of the same manufacturing cycle, which means they were all made at the same time. It turns out, of the thirty devices that were manufactured, all had the code present, but it was not activated.

Just think about that. When you utilize systems in high-risk areas, whether it's in hospitals, nuclear reactors, air traffic control, and others, these critical systems go through extensive testing for years and years before they're ever deployed. This is an example where thirty different systems were all inspected, verified, and validated by thirty different entities, looked at for over three to four years, and not one of them was able to find the malicious code. If we weren't able to proactively detect malicious code in a system that runs a nuclear reactor, what are the chances that we're going to find it in our laptop, tablet, or iPhone?

The other interesting thing about this story is when Stuxnet occurred it turns out that malicious code was in that device for nine years running in a production environment, and once again, nobody caught it. If you're having a device run for over eight years and nobody catches the malicious code, this just shows you not only how good the attackers are, but how complex these systems are.

THE CHALLENGES: AN UNFAIR GAME

Verifying every single component and functionality while finding any potentially malicious code is extremely difficult to do, but hiding the malicious code is relatively easy. It's not that the offense is necessarily better or smarter than the defense, it's that they have an easier job.

In a fair game, I will take our cybersecurity professionals against any country's offensive capability. In a fair game, we will have a chance of beating them every day of the week. But this is not a fair game. If you're an attacker and you're playing cyber offense where you're trying to break into a system, company, or network, how many vulnerabilities do you need to find? Just one. That's it. Just one vulnerability is all that's required. Now, if you're on the defense and you want to protect and secure against attacks, how many vulnerabilities do you have to find? All of them. Not 90 percent, but all of them. Because if you only find 90 percent, 10 percent still get through, and the offense only needs one vulnerable point.

The trick is not to go in and stop them from scoring, but to be able to detect them in a timely manner when they do score. The fundamental problem we have in cybersecurity today is that organizations, entities, and even the country believe that you can prevent all attacks, and that's a losing game.

One of the things that I struggle with is that the main cyber-security approach of the United States is heavily focused on preventing and stopping attacks. Don't get me wrong, there are some detection capabilities, but the mindset we need to all start adopting is to prevent what we can but put a heavier emphasis on detection. I want to be careful here because there are still a lot of things that you can block, so that should be part of it. Prevention is ideal, but detection is a must. You can't prevent all attacks, but you prevent what you can and then detect the rest.

The other challenge that we run up against, as I've said, is the media. They don't really cover cyberattacks. Sometimes if you search really hard, you can find the story on a second- or third-rate news channel or online, but the main news channels are not covering this. Some may, once in a while, if there's something really big, but we know big breaches are happening often, and they're just not reporting it. Fighting between Democrats and

Republicans seems to be a much more interesting story to the press and the public than cyberattacks. Cyberattacks tend to be more covert and harder for people to understand. If a politician sends an obnoxious tweet, that is easy to understand and hard to deny. If 100,000 passwords were compromised, it is not only harder to comprehend, but harder to anticipate the direct impact.

The real solution is that we need to help the media, not criticize them, because they are covering what they know. The more we can educate the media, and the more information we can provide them, the more they will run stories that impact citizens around the world.

OUR CRITICAL INFRASTRUCTURE

Now, when we look at critical infrastructure, what we're really talking about is what we call ICS (Industrial Control Systems) or SCADA (Supervisory Control and Data Acquisition) systems. These are essentially the components that run the critical infrastructure, like the devices that create electricity, control the cleansing and flow of water, manage and run nuclear reactors, handle air traffic control, and run life-support systems in hospitals.

These systems are designed to be very reliable but also verified and validated as simply as possible. That means these systems do not have a lot of complexity or error checking, and (I know this sounds crazy, but I'll explain why) not a whole lot of built-in security. The reason is simple: These devices were designed and built to be deployed in what we call an air-gapped environment, meaning it's isolated and not connected to any other network.

In a nuclear reactor, these ICS systems that run the reactor were meant to be on a separate, isolated network, not connected to any other network or any other component, which means most of the security was based on physical protection and parameters.

Most nuclear reactors have pretty good physical security. They have armed guards, fences, monitoring. Good luck if you think you're going to get into that reactor. The expectations of these ICS systems was that somebody could only do harm by getting physical access to the device, but because they are on an isolated air-gap network, that would be extremely difficult to do. So it was deemed as an acceptable level of risk.

The reason these systems were built without a great deal of security complexity was that the designers wanted them to be straightforward, verifiable, and reliable. They wanted these to be able to run for twenty or thirty years without any updates or the need for new components. The more complexity, error checking, and security you add in, the more lines of code that are created, and the greater the chance that vulnerabilities, exposures, or errors may be missed or induced.

That simplicity worked well as long as companies were honoring the isolation and air gap of these devices. What started happening over the last three years, which is terrifying to me, is for functionality purposes, the companies started connecting these devices to other networks. I've even seen situations where these critical systems were actually connected either directly or indirectly to the internet, which now becomes incredibly dangerous. These devices control very critical functionality and were built without any security in place. If you put these systems out on other networks or the internet, it is easy for attackers to potentially break in.

The scariest part isn't about sensitive data potentially getting compromised, but the ability for someone to crash these devices and start taking down critical infrastructures like water and power distribution systems, etc. Typically, the easiest type of attack goes against availability. Breaking into a device to steal and/or alter information requires access, authentication, and other critical skills.

But to crash a device, take down a device, or make a device nonfunctional, all you have to do in some cases is just run basic vulnerability scans against those systems, and they will go down fast. Then we start getting into mass chaos and things that could very easily cause loss of life. One of the tools we use in cybersecurity is called a vulnerability scanner. These are publicly available on the internet. There are free versions or trial versions, or you can even buy a full commercial version as a contractor for $2,000 for an annual subscription. These tools look for vulnerabilities on devices. If you run this against your computers in a normal business, the scans will work fine and come back identifying any potential vulnerabilities that might be present.

But here's the kicker: If you take those same vulnerability scanners and you now run them against industrial control systems, PLCs, or other components that manage water, gas, nuclear, or oil, with their simplistic setup, when you're sending unusual packets and things that are meant to test and verify and validate, they will crash, because they don't have any of the error checking. If you run most vulnerability scanners, even free ones, against an industrial control system, they will take down the environment quickly.

If you work in infrastructure or are just a citizen of the United States, you need to recognize that what we're doing right now is dangerous. We're taking critical systems that were never meant to be connected and making them directly accessible from the internet.

There was a large chemical processing company that made a variety of chemicals including chlorine. If chlorine gets exposed or is not processed correctly, it can cause death. This company had a business network, and an operational industrial-control network. These networks were air gapped, and from the business network you could not access the operation network. So instead of doing the right thing, which would require setting up access with one-way connections to allow functionality but to limit exposure, they had one of their contractors set up a portal from the internet that, with solely a user ID and password, you could access the entire operational industrial-control system network.

You know by now that figuring out a user ID is not hard. Brute forcing passwords is not difficult. Their system didn't require users to change their passwords at all. Some people had the same password for two years. To make matters worse, nobody was monitoring it, so if somebody was actually trying to hack a password, the business would never know.

The idea of security through obscurity is believing that if the inner workings of the system are hidden, like putting a system out on the internet that nobody knows about or understands how it works, that it is secure by default. That doesn't fly. Attackers today are constantly scanning, and they have automated scripts where they can go through the entire range of IP addresses within thirty-six hours. Any server that has access from the internet has to have a public IP address, meaning it can be found and scanned in less than two days. Security through obscurity doesn't exist.

Still, there was pushback at the chemical processing company, so we set up logging and alerting, and within a week, we found that there were more than ten accounts being brute forced. In two of the cases, the attacks were actually successful. This system had been out there for eight months prior, so it could have happened more times than that. This was just over the course of one week that we were watching and saw two accounts compromised and eight accounts targeted.

Looking at critical infrastructure and targeted attacks and our nation as a whole, we can make some general assertions based on what the attack is doing and where it is coming from. If the attack is stealing intellectual property, it would usually be coming from China. If it was going after financial data, it would normally be originating in Russian. And if it was causing disruption or taking systems or devices down, it would typically be from Iran or North Korea.

However, this is starting to change. We're beginning to see many more countries launch cyber-based attacks for monetary gain. As you'll see in the "Finance" section of this chapter, North Korea is struggling from a monetary perspective, so they now have an entire group whose sole purpose is to try to break in and steal personal information and financial data to sell on the dark web or to utilize for monetary gain. We'll start to see a lot more countries do that. Also, China is now recognizing that one way to have a competitive advantage is to steal intellectual property. Another way to have a competitive advantage is to take down manufacturing and impact that sector's ability to produce. For example, if a US entity produces 40 percent less, they're going to have to buy more from overseas entities like China, which gives China an instant competitive advantage. In the last couple of months, things have changed dramatically in terms of who's attacking, what they're going after, and who's trying to cause harm.

Two years ago, we were monitoring attacks from countries into the United States, and as mentioned, we saw a lot of attacks from Russia, China, and North Korea, but we also saw a pretty good amount of attacks from countries that we would consider allies, like the UK and Canada.

But the thing that stuck out in my mind was that we didn't really see a lot of attacks from Iran. We are definitely not considered the best of friends, so you would expect a fairly good number of attacks coming from Iran into the United States. It was almost too few, to the point where we questioned it. It took us a while, but what we realized is that 15 percent of the attacks that were coming from China were actually Iranian. The Iranians have very advanced cyber capabilities, and recognized that if they attack the United States directly, that's poking the bear. It's going to cause retaliation; it's not a smart thing to do. So they broke into a lot of vulnerable systems in China, and then used Chinese IP addresses to break into the United States.

Which brings us to attribution. I (and you) need to be careful when we say things like 80 percent of attacks are coming from China. The reality is that 80 percent of attacks are coming from IP addresses in China, but that doesn't mean that the attacker is sitting in China. That's just the last stop, and they could be anywhere else in the world. It's fairly easy to move and relay packets.

The second example is North Korea. Not to get too technical, but there are different protocols available today, and the backbone of the internet runs what we call IP internet protocol version four, known as IPv4; however, some countries can also run internet protocol version six (known as IPv6), and the interesting thing is that they're not compatible. So if you're going to go from IPv4 to IPv6, you need to do something called tunneling. And if you want to go from six to four, you need to do the translation. You actually have to explicitly convert back and forth.

One of the things that North Korea did that's pretty clever is that the backbone of the internet is IPv4 and their internal network is IPv6 (the v refers to the version numbers), so they have four entry points into their country and six entry points out (it's just a coincidence that the version numbers match the number of points). When North Korea launches cyberattacks, one of the things they do is shut down the inbound connection points, which means they can now fire packets outbound, but we can't attack back. Other than a physical attack of dropping bombs, there's little we can do to stop them when the North Koreans attack us. I hate to say this as a US citizen, but it is pretty clever in terms of how they operate.

That's probably the United States' biggest disadvantage. Most other countries know and control their entry and exit points to the internet, so if they want to shut down inbound connections and/or outbound connections, they absolutely can. Knowing a country's connection points to the internet is a very strategic asset. Just this year, Russia was able to completely disconnect from the internet for a day and not allow any packets in or out. Unfortunately, as I mentioned before, because the United States built the backbone of the internet, this is not something that we can currently do.

Earlier, I said the fact that the Russians are in our critical infrastructure and we're in their critical infrastructure is sort of like the Cold War, and if either side does something, the other will retaliate. But what if the Russians decide to attack and take down our critical infrastructure, and then they disconnect from the internet? Then we can't launch a retaliatory cyberattack on them. Then what?

So far we have seen cyber versus cyber, where a country has launched a cyberattack against the United States and we have launched a cyberattack back. We have also seen cases where somebody has physically attacked a country and that country retaliated with a cyberattack, as with Iran. What we haven't seen yet, which I predict we will see very soon, is somebody launching a cyberattack, and we retaliate with a physical attack. That is a hard one for many people to understand or grasp. If you send me packets, does that ever justify dropping bombs on you? Most people would say no. But what if those packets alter the water supply and poison and kill twenty thousand people? And then you disconnect from the internet so I can't attack back in a cyber manner. That might justify physical retaliation against that country, right?

The bigger issue that I mentioned earlier is attribution. How do we know that it really came from that country? What if China broke into Russian servers and launched an attack against the United States that took down the power grid or poisoned the water supply and caused loss of life, we trace it back to Russian servers, and we bomb Russia? Right now we do not have the capability to do full traceback and verification. A third party could cause a lot of problems . . .

EVERYDAY INFRASTRUCTURE

Most of us don't really think very much about how dependent everyday services are on sophisticated computer networks. You're probably beginning to see it with healthcare and computerized medical records, but it pervades just about everything in our lives from stoplights to 911 dispatching, store inventories to airplane flight control—all of it is in danger of being brought down.

Healthcare

You could argue that if intellectual property or manufacturing data is exposed or given to a third party, that could hurt a company financially and put them out of business. But when you're talking about healthcare information, hospitals, and doctors, if critical medical data is altered, if people are given the wrong medication or treatment, if ICU equipment or life support machines are overwhelmed on a network and fail, that could be the difference between life and death.

First and foremost, when people think of healthcare and cybersecurity, what immediately comes to mind is HIPAA, the Health Insurance Portability and Accountability Act. This focuses on confidentiality: making sure that your information is properly protected, secured, and only shared with other doctors or medical professionals whom you approve, authorize, or allow.

You could argue in healthcare that integrity and availability are probably more important than confidentiality. If somebody finds out that you have a certain illness, you might be concerned that they would discriminate against you at work or not give you a promotion. But does somebody knowing about your medical data directly put you at risk? What can have more impact on your life is if somebody can alter or modify your information so you're given the wrong treatment, or your information is not available to attending physicians, or the medication is not available when you need it.

There have already been several factual documented cases of cyber hits where criminals have targeted people within hospitals and have actually killed them by altering medications, modifying dosages, or taking down equipment. Just like in the physical world, criminal elements perform "hits" in which they kill key individuals to either send a message or to eliminate people who are not cooperating with them. These entities are now using cyber technology to accomplish this goal. This also further emphasizes my point that many crimes are moving to cyberspace because it is easier to perform, and in the case of murder, there is plausible deniability. With a traditional hit, you have to physically approach the person, use a weapon, and flee the scene. Since this is not an episode of *Law & Order*, I will leave out the gory details. However, with a cyber murder, someone would alter or modify equipment to make it easier to perform the deed. As I am doing the final editing of this book, there was an episode of *NCIS* in which someone on a navy ship broke into the computer of the car of someone she was having an affair with. She remotely disabled the brakes and drove the car into a tree, killing the spouse of her lover. We are seeing things like this happening in the real world.

HIPAA is so focused on confidentiality that in some instances it negatively impacts the availability of information. I know several cases where somebody had to go in for emergency surgery and, because of HIPAA, the doctors or the hospital that they went to were not able to get timely and quick access to their vital information. People believe HIPAA was created as a cybersecurity law to protect us, but in some cases, it has the opposite effect of actually reducing or limiting access to data that healthcare providers require.

Let's look further into some of the challenges we have in the healthcare arena. As I said, it's important for everyone to understand these challenges, whether you work in healthcare, manage a business that provides healthcare for employees, or simply to safeguard your personal information.

The typical tech refresh cycle is anywhere between three to five years. That means in most companies you would get a new laptop every three to five years, and the servers would be replaced and updated. Most companies use that opportunity as a time to upgrade the operating system. If you look at most of your major operating system vendors, they typically keep an operating system supported for at least eight to ten years. So if you're doing tech refresh every three to five, and you tie that to the operating system upgrade, that turns out to be a nice solution. You do still need to make sure you're patching and updating the software during that period, but the major operating system upgrade is only done when you roll out new hardware.

When it comes to the medical industry, because this equipment is dealing with an actual human patient, there could be loss of life if it malfunctions. The high-end, expensive equipment is designed to last twenty to twenty-five years and takes three to five years to test before it's even rolled out. So it's not uncommon for hospitals or doctors' offices to still have outdated operating systems like XP or Windows 2000. Today, most operating systems become obsolete in fifteen to twenty years. That means the primary challenge in the medical setting is that a lot of equipment is running older, outdated operating systems because of stability and uptime availability.

Before deploying these systems, manufacturers and organizations using the equipment are going to test and make sure these systems are highly stable. If an outdated operating system like XP or even 2000 is left by itself, not connected to any other network and properly maintained, it is actually still very stable. It could be up and running for eighteen years with no crashes—that's a pretty stable operating system. So trying to update operating systems every six or eight years is very dangerous because whenever you alter, modify, or change any component, the probability of something crashing or not working is fairly high. Many of these systems in hospitals and medical offices are not updated. They're not patched because they have to go through testing to make sure they're solid, stable, and working correctly.

Here's the interesting twist: They were never intended to be connected to a network. Most of this equipment was meant to be standalone, isolated, or implemented in air gaps. Air gaps are probably one of the best security techniques out there. As we covered earlier, an air gap system or an air gap network is exactly what it sounds like—it is an isolated network that is not connected to any other network, especially the internet. With a network that is air gapped, unless you can gain physical access to the building and connect directly to the network, it is not accessible from other networks. So it is a medical device that has a computer in it that is sitting in a hospital, not connected to any other system, any other network, and the only possible way you can get to it is by physical access.

A lot of these devices were built and maintained based on the premise of isolation and really good physical security. Even though we are discussing medical technology, we see very similar parallels in the utility business dealing with nuclear reactors and other very sensitive equipment. Once again, all of this equipment, whether it's running life support in an ICU or it's running a nuclear reactor, was never meant to be connected to a network. It was always meant to be behind very heavily guarded, very secure areas and facilities that would be very difficult for an unauthorized person to physically access. They were never, ever meant to be connected to a network.

This is an area where you have to be extra careful with cybersecurity. Traditional cybersecurity measures do not apply to specialized environments like hospitals and healthcare. The way many security companies perform assessments is to gain access to the network and perhaps even a wireless network. They might try to connect either the old RJ45 Ethernet jacks, which are becoming less and less common, in a conference room or even at a public reception area, or just directly connect to an open wireless network. From there, they scan the entire network and try to access every computer on it. They do what we call an asset inventory or host discovery. Once they find those computers, they do what we call a port scan to find out all the areas that you can access the device and all the listening components. They'll try to access each of those areas, looking for any vulnerabilities or exposures.

If you're dealing with a traditional company or organization that's running Windows, Linux, or Mac operating systems, in most cases, it's fairly safe. That's because those are very robust operating systems, and they have error checking and other measures in place. If you send them unusual packets or information they're not expecting, or if you try to break in or compromise the system, they have the proper mechanisms to be able to handle it and continue to function.

The problem is, in the medical industry, you want those devices as simple as possible. The more complex a system is, the more code that is running, the more things that can go wrong, the more testing and validation. So many of these medical devices, because they were designed to be air gapped and meant to be simple and fully verified from a testing perspective, have very simple operating systems. They're expecting a packet configured in a certain way, and if you send it exactly the way it's expecting, it processes it; however, if you start sending packets, data, or information that the device is not expecting, it doesn't know how to handle it. Unlike other general business systems, if these specialized devices are running an operating system like Windows, they'll typically crash and go down when the systems are scanned. So you want to make sure that you're very careful with who you bring in to do security.

When my company bids on work for the medical industry, which we do often, clients may ask me why my quote is much higher than my competitors.

I'll respond, "The other people that you got quotes from, do they have extensive experience with medical and hospitals? Do they understand the intricacy of the environments?"

Most of the time they'll reply, "Well, no. They said they're security professionals, and all industries are exactly the same."

I'll advise: "Don't sign the contract. They don't know what they're doing. They're going to do automated scans, which are cheaper, but they're going to be detrimental to what's going on, and it's going to put you, the administration, and everyone in that hospital at risk."

When it comes to cybersecurity, you're always going to pay the piper. You either pay now or pay later. We provided a bid to one Midwestern hospital for a massive project, which they ended up giving to a cheaper competitor to save some money.

I had warned them: "Listen, you can hire somebody else and save $20,000 on the proposal, but when they crash your systems and when things don't work correctly, it's going to cost you probably a few million dollars to recover and get back up and running. So it's up to you. Do you want to spend $20,000 more now or do you want to spend $1 million later?"

It's always going to be that way. Recognize if you're cutting corners or cutting costs; always ask yourself: Do we pay now or do we pay later? Trust me, paying later is always more expensive.

This particular client didn't listen to me and chose the cheaper consultant. A few months passed. It was a beautiful Saturday in May and I had just gotten back from four weeks on the road. I travel a lot so I was really enjoying being back with my family. I was in the yard with my kids, and we were swimming in the pool, steaks on the barbeque.

I was relaxing when my phone rang: it was the Midwestern hospital. Before I could barely even say hello, I heard: "Eric, pack your bags, head to the airport, book the next flight to Chicago, and call us on your way and we'll brief you on what's happening."

I tried to get more information before leaving for the airport, and we went round and round on the phone. Finally, they said, "We hired this other company and don't say you told us so, but they went into the emergency room acting like a patient and they found a network jack on the wall. They plugged into it, and they had access to the entire hospital, and they started doing automated scans across every device in the hospital, and they crashed all of the life support systems in the ICU. They crashed all the MRI systems, and they not only put patients' lives at risk, but they paralyzed the entire hospital."

In that particular case, unfortunately, as much as I would've liked to stay home, I was needed to help out the client. They were compromised and needed a professional. One of my many specialties is that, when there are major breaches and a director, CEO, or president is very upset, I'm the negotiator the company or CEO hires to help explain what happened and to calm the situation. I joke that I'm a marriage counselor because I can interface between technical and business and executive and strategic.

Truthfully, in that situation, there was a lot of blame to go around. First of all, they never should have hired a company that had never worked in the medical environment before. Second, that company never should have scanned the entire network. I will tell you, even with stable networks, we *never* scan the entire network. That is high-risk poker. If you scan an entire network and there is an exploit that causes a system to crash, all of the systems crash, and the entire organization is impacted. If an organization has a thousand systems and you scan all of them and they all crash, that has a huge impact; however, if you only scan five of the thousand and they crash, you can typically recover.

Continuing with the example, they also had a network jack in the emergency room. A public, open area never should have access to the rest of the network, and the MRI and the ICU systems never should have been open to the rest of the network. And the list goes on. When you're dealing with healthcare, there are unique challenges that you need to be aware of.

What is happening today in the healthcare industry is that organizations are trying to be more productive and make things easier for the staff: One way of doing this is to connect medical equipment to the business networks or other networks. This is never a good idea. Every hospital I know of that has connected life support, MRIs, and other critical equipment to an open network starts looking at the value and benefit and sees the risk, exposures, and issues, and they always roll back the decision. In my opinion, there's never a strong enough justification to allow systems that don't have proper validation, error checking, and cannot handle basic scans to be connected to an open public network. It's too risky.

Another thing to keep in mind is to make sure you're addressing all three components of the CIA triad. Most hospitals and healthcare agencies today are doing a pretty good job on the confidentiality side because of HIPAA. However, they are not doing a great job with integrity and availability. When most healthcare organizations consider availability, they think of what we call a transparent backup, meaning that they replicate the data in real time. So if there is some accidental occurrence—let's say that a server crashes, or a system goes down, or there's a hardware failure and that data is no longer accessible off that system—it can be recovered from another device.

A great example of this is the WannaCry attack I referred to in Chapter 1, which happened to three interconnected hospitals in the UK. You would think they would have had triple redundancy, so that if any one system or one hardware component failed in any hospital, the data could easily be recovered from another system.

Even if two devices simultaneously failed, which is highly unlikely, they would still have a third component in place. From a traditional backup standpoint, this actually seemed to work quite well.

Ransomware attacks changed everything because, in that case, the modified data replicated to all three areas. At those hospitals, one high-end doctor, who had access to 80 percent of the patients' records, clicked on a link. At *his* hospital, all that data became encrypted and inaccessible because that's the whole idea of ransomware—unless you pay the ransom, you won't get the data back.

What do you think happened next? The system thought there was a data update, so it automatically replicated to the two other hospitals. Within fifteen minutes, 80 percent of the data across the three hospitals was encrypted and unrecoverable unless they decided to pay that ransom. They had redundancy—but it didn't work the way they thought, and it certainly didn't work in their favor.

There is a difference between transparent and nontransparent backups. A transparent backup is one that automatically gets replicated. In a nontransparent backup, even if the data changes in one spot, it doesn't change in the other. This shows that even though tape backups might seem old-school, they still have value in some settings.

Remember, you are a target and cybersecurity is your responsibility; it's critical for you to remember this because all your medical records—in your doctors' offices, in any hospitals you have visited, in labs—have your social security number, your driver's license information, and other important data.

Here's the kicker: Many states and many countries don't require them to have that information. They'll ask for it on the form, and if you voluntarily provide it, they're going to be more than happy to take it. Be careful when you're filling out any forms with sensitive information, and first ask:

Is this data really needed/required?
What do I risk by providing it?

I'll keep saying it; always ask the question:
What is the value and benefit of providing my social security number versus the risk and exposure, and is that risk and exposure worth that value and benefit?

I'd argue there's no value. In many situations, they will ask for your social security number on the forms, but you are not actually required to write it down.

There are two final things I want you to consider about your medical data. First, how is it protected and secured? A medical facility should provide you with a general statement about security and privacy and how they are protecting your information. The second thing is that you can ask doctors and hospitals to remove your information. They keep it for five years, but you can override that if you move or change doctors. Get the information and maintain your own records.

Finance

When many people think about cybersecurity and areas that can impact their life, their business, and their family, finance and healthcare are always at the top of that list. I will tell you financial security needs to be a top priority. More and more cybercrime is financially driven. When most people think of cybercriminals, they think of governments trying to steal top-secret information, proprietary data, or critical intellectual property. While some of that is still happening, what many people don't realize is most of the attacks we're seeing today are actually financially driven.

Many of the cybercriminals in Russia have formed companies that are focused on making money from cyberattacks: bank account break-ins, fraudulent transactions, identity theft—anything that can drive monetary value.

We've recently seen that North Korea is actually utilizing cybercrime as a source of financial funding for their country. Building rockets is expensive, and North Korea is getting low on cash. Instead of targeting the United States for disruption like the North Koreans are typically known for, they're committing more financially driven attacks, going after bank accounts and other financial records. They are actually giving the Russians a run for their money. Literally.

You need to make sure that you're focusing on proper financial security and understand the ins and outs in order to protect yourself, your business, and your family. First and foremost, there is a big difference between debit cards and credit cards. I know financial advisors like Suze Orman and many other folks actually push debit cards because they tap into the actual amount of money you have in the bank, so you can better manage your funds, instead of running up debt on a credit card.

Unfortunately, their reason for pushing debit cards is the exact reason why I say they should be shredded and nobody should have a debit card in their wallet. If somebody steals it and commits fraud, you're out the money. If somebody withdraws $200 on your debit card, it immediately comes out of your account.

Depending on what happened, if you weren't at fault, you could spend time and argue with the bank, proving to the bank that it was a fraud. And it could take two, three, or I've seen some cases that take up to nine months. I've even seen some instances where the bank said, "You didn't protect your account, you were phished, you gave it up; therefore, you're liable." Because debit cards come directly out of your account, that makes you weak in terms of negotiating with the bank.

Let's now talk about credit cards. First of all, with credit cards, regardless of whether you are at fault or not, the money doesn't come out of your savings or checking account and instead is billed to the credit card company. When you report a fraudulent transaction, you're not out the money. Even if it takes the credit card company six months to investigate, they are out the money. It puts you in a safer position.

Second, Congress passed a law many years ago that said individuals are only liable up to fifty dollars on credit card fraud, pushing all the liability on the credit card company. This also puts you in a better position because in the worst-case scenario, it costs you fifty dollars—that's it. That protection doesn't exist on debit cards.

Even if you don't listen to me and you use debit cards and/or credit cards, in both cases, you want to set text notifications to alert you when a charge is made. I know it sounds like a lot of trouble to go through, but as long as you approve the charge within thirty seconds, the transaction is approved. It comes over text and it just requires a simple reply.

If it's a fraudulent transaction and I'm in a meeting, and more than thirty seconds pass and I ignore it, guess what? The charge is denied. You're in control. Even with credit cards, if somebody does a fraudulent credit card transaction, you're only in a position of strength if you detect it.

When we talk about credit card fraud, you probably think it's going to be a big number. You scan your credit card statement looking for an outlandish amount. What you don't realize is that most credit card scams are skims, which means they are about three to five dollars. Most people never notice. The big reason you want to set up the text notification is that most of us don't keep our receipts.

One of the big areas of attacks is skimming on the tip. When you get the initial bill at a restaurant, they do a pre-run of your card, and then you add the tip and leave. You don't get the final receipt. So if you tip twenty dollars, but the waitstaff enters thirty dollars, unless you get verification on the spot, you probably wouldn't know or remember. I know people who have lost thousands of dollars because every month it was two dollars here, four dollars there, and nobody notices at the time—until we went back and checked. If you do the text notification, it's easy to notice. Now if you live in a small town with a few restaurants, and you are friends with the wait staff, this might sound surprising, but in larger cities and bigger restaurants, this happens on a regular basis.

This just happened to me the other day: I gave a fifteen-dollar tip, and I received the text notification for thirty-five dollars in tip. I was still at the restaurant, so I called the manager over and explained what happened. And of course, he said this sometimes "accidentally" happens—and he actually did the air quotes. That's the problem that you're up against. I know you might think it's difficult to verify each transaction, but you're going to pay the price—either pay now or pay later. You can either spend two additional seconds verifying your transactions when you make them, or you can lose hundreds or thousands of dollars.

Let's move on to all-out financial attacks. Probably the biggest one—I get at least two or three of these every single week from corporations—is fraudulent phishing emails. Remember, these attackers are smart, and phishing is popular. They know your business; they know your customers. In many cases they have access to and they're monitoring your emails. So this type of scam can look a couple of different ways.

One is that you'll get an email from a customer that says something like, "We want you to make this month's payment to a new account. We're setting up some new financial accounts to diversify our interest, and we would like you to send the funds to the following account."

This happens all the time. Vendors that any company makes monthly payments to are always changing and updating accounts. It is not an unusual email. Some businesses or clients are getting four or five of these emails a day.

Normal business practice is to send back an email to the person in charge and say, "Hey, we just want to verify this as a correct request, and we need your verification that we can change the account."

Then the business will receive a response saying, "Yes, we did. Thank you for checking with us. We really appreciate you verifying it to avoid any fraud on our account. But yes, this is approved and this is a new account."

You might remember what happens next from a similar story I told in Chapter 1. The problem is the attackers are in the email system, so they're spoofing and sending back communication.

Here's where this gets particularly challenging: Also as mentioned, when it comes to these fraudulent transfers, you typically have twenty-four hours, but most of these accounts have so much high volume, businesses don't notice until the end of the month. Your client will come to you twenty days later, asking for the money, and you'll be confused, thinking you paid it, only to review it with them and discover it wasn't their account. And because your staff actually transferred the money, you're liable. It's not the bank's fault.

Once again, you pay now or pay later. The trick for combatting this is you want to do out-of-band verification. Either pick up the phone, get a physical signature, or some other secondary method. If you're getting a request on one communication channel, always use a different communication channel to verify.

As mentioned earlier, I was trying to fix this particular problem with a company that was out a half million dollars. I told them anytime there's an email to do a financial transfer, the CFO needs to send the CEO a text to verify. The CEO frowned.

"But Eric, that's going to take an extra five to ten minutes a day because it's going to take thirty seconds, and I do ten or twenty of these a day, so it's going to take some more time."

I said, "Okay, sir, which is better: a couple of minutes a day or four and a half million dollars? Let me ask it another way: Is a couple of minutes a day over a three-year period less or more than four million?"

Obviously, it's far less. He finally understood.

It goes back to one of my early lessons: With cybersecurity, you're going to have an inconvenience. Do you want to be shot in the leg or shot in the chest? Now I know most of us don't want to be shot, but that's not an option. It's going to be a little work up front or a lot of work on the back end.

The cousin to phishing is caller ID spoofing. You'll get a phone call, and the caller ID says it's your bank.

They'll say, "We're seeing some unusual activity on your account, and we need to check and verify that your account is valid. We're going to send you a text right now with the code, and we need you to repeat the code to us so we can verify and validate that you're authentic."

They're acting like they're helping you with your security when in reality, they logged into your account with a password, and because it came from a different location, the bank needed to authorize it with a one-time code. They're trying to log in, so with you on the phone, they hit submit and you get that text. You read it to them and now they're able to log into your account. They say thank you, hang up, and wipe the money out of your account.

Never give out passwords, PINs, or one-time information over the phone because we're starting to see some low-tech attacks like this that are very effective.

One of the other solutions that has been around for a while is to freeze your credit. This was popular around the Equifax breach mentioned in the Introduction. If you're not buying and selling houses on a daily basis, you need to freeze your credit. This is as simple as calling the credit agencies or, if you are tight for time, you can pay a small fee and have a company do it for you. That stops people from pulling information, opening up lines of credit, and committing other financial types of crime.

The next thing you can do is look at the security for all your financial institutions. Many of them have great security functions but they are not turned on by default. You want to go into the websites of your bank, credit card company, etc., and—you guessed it—turn on two-factor authentication and account notifications. This way, any time somebody tries to log into your account, transfer money, or perform any financial types of activities, you are notified immediately and can go in and take proper action.

Don't think for a moment you'll stop all attacks. You are going to be compromised and targeted, and what you need to focus on is timely detection. The more visibility, alerting, and awareness you have of your accounts, the better you can do to protect and secure your financial information.

CYBERSECURITY, ELECTIONS, AND NATIONAL PRIORITY

More and more states are utilizing e-voting. E-voting is not secure, so every voting place in the country that wants to have an e-voting option has to offer both options—paper and electronic. You can argue that paper voting is more susceptible to mistakes through human error, but with paper ballots, you can recount and revalidate. You have checks and balances. But if I hack into the e-voting system and change people's votes, you can never validate. Computers make fewer mistakes than people, but they can't validate. Because manual vote counting takes more poll workers, most places might have only one paper ballot and fifteen e-voting systems.

When I went to vote in the last election, they said it was a twenty-minute wait to fill out a paper ballot, or I could vote immediately electronically. So even if, by law, they give both options, they don't have to be equally convenient. There have been numerous documented cases where these systems were not validated to be secure and were connected to other networks. This allowed them to be accessible to attackers and potentially compromise the information. Functionality leads to security problems. These systems are vulnerable and yet, they are still being utilized, and even more so in future elections.

The US government is looking at the wrong problem. It's not collusion . . . it's voter influence. The last couple of presidential elections were extremely close, and that's the scary part. Hackers wouldn't have to manipulate many votes. We did the analysis of the last election, and you would have only needed to have broken into three voting locations in Pennsylvania and two in Florida to skew the results. And who runs and maintains these voting systems? Senior citizens who didn't grow up with this technology and yet are tasked with setting them up.

You can wait in line to vote, but we as American citizens have to recognize that we do have a say in running our country, and we influence big issues. If you look at controversial issues, they have shown over time the American public has influenced them by voting. Cybersecurity is the next big issue that people need to recognize that affects them; start engaging in it, and start making a difference.

The problem with cybersecurity is that it was always one of the top three priorities for the previous three presidents, but it's never been number one. While it is good that the directors of our country understand and recognize cybersecurity as an issue, it never makes it to the point where serious action is taken.

While listening to the primaries for the Democratic Party in 2019, not one candidate mentioned anything about cybersecurity. No one talked about the importance and criticality of cybersecurity in any of the debates or speeches. Until we make it the number one priority, we're going to continue to have problems.

For almost four years, you heard talk about Russian intervention in the elections and whether they interfered or colluded. The bottom line is that we've known for over ten years that our e-voting systems are vulnerable. We've known for many years that electronic voting systems should never be accessible from the internet, yet almost a third of states keep their e-voting systems up and accessible from the internet 24/7, 365 days, and never check or fully verify or validate for cybersecurity attacks. There was a lot of squabble about whether it happened and who was at fault, but no talk about fixing anything.

We need to stop trying to treat the symptoms, and when we see a real problem, we have to fix the root cause. The first thing that must be done is to recognize where the vulnerabilities are and remove those from the network. With e-voting, any system that's been accessible from the internet should be decommissioned immediately. Even if the system is online for forty-eight hours, the probability of compromise is extremely high. And with the amount of energy and effort it takes to verify or validate one of these systems, it would be much cheaper and more reliable just to replace it. Why would you spend $10,000 for security experts to verify and validate whether it's been compromised—which isn't 100 percent accurate—when you can just use a new system (that is close to the same price) that is secure and locked down?

Second, e-voting systems should have strict standards. We have the NRC—the Nuclear Regulatory Commission—that oversees both physical and cybersecurity for all critical infrastructure relating to nuclear reactors. Why isn't there anyone focusing on the cybersecurity of our e-voting systems? Why aren't there standards for locking down, securing, verifying, and testing? And if people have proven that these systems are vulnerable and exposed, why aren't these problems being fixed?

Right now, writing this book, I would recommend that all states move back to a paper system. However, with the spread of COVID-19, it might make more sense for the federal government and states to invest in electronic options for voting that are more secure and properly validated.

Yes, you could argue that paper has its issues, but with a paper voting system, you can verify and validate that a ballot hasn't been altered or modified. It's a piece of paper that somebody fills out and, if needed, you can go back and revalidate and recalculate the votes. With electronic voting, you don't know if the votes have been altered, modified, or even deleted. When you look at some of the numbers from previous elections that have utilized e-voting systems, you get some really unusual scenarios, like three thousand people voting, yet there are five thousand votes. Or, the other way around, three thousand people voted and there were only one thousand votes, and if you look at how close some of these regional, state, and even national elections are, every vote counts.

Even if somebody could go in and manipulate only 5 percent of the e-voting systems, that could significantly change the outcome of who gets elected and who becomes president. And remember, that has all sorts of ramifications. If we think, for the moment, only about presidential elections, the president names Supreme Court justices, federal judges, the heads of many federal agencies, hugely influences foreign policy . . . now follow that downstream, and it also influences senators, representatives, your state and local officials. These are the people that essentially make the policies that control our lives. When you're looking at something as critical and as serious as e-voting, we need to focus more attention on securing, locking down, and protecting those systems. It has been proven time and time again that the current systems have not been secure and/or have been compromised.

We need to start over, put strict policies in place and—until we validate that security—verify that if three thousand people vote, there should be three thousand votes in the system and prove that it hasn't been altered, modified, or changed. We need to stick to paper until we're ready for technology. E-voting is another perfect example of how people do not think they're a target, and most people think about functionality first and security second.

No one wants to admit that they're vulnerable, and nobody wants to admit that cybersecurity is their responsibility. Instead, we keep running away from the problem. Whether it concerns e-voting, or your business, or another area, you are part of the infrastructure of the United States and you need to start protecting and securing your critical information.

NATIONAL REGULATIONS AND YOUR BUSINESS

National regulations are something that could impact your business at any time. Currently, the United States is one of the few countries that does not have national laws on privacy. We do have various state laws that typically contradict each other. For the most part, California is usually the strictest, so we usually recommend following that, but it's also good to look at the global arena.

Over two and a half years ago, the European Union passed GDPR, the General Data Protection Regulation, which impacts any European citizen. Anybody who is a citizen of the EU is covered under GDPR in terms of limiting the collection of information, making sure it's only used for the purpose that's required, and that the data is properly protected, secured, and locked down. Probably one of the harder ones for most companies is what they call the right to be forgotten. If you contact a company and you say, "I want you to forget me, I want you to delete any information you have on me," the company has to delete the information and completely remove it from all of their systems. That might sound simple, but it concerns more than the company's active database—there are generally backups and backups of those backups, so removing all that information might be very difficult.

Technically, GDPR only covers citizens of the European Union, but what if you're a local doctor's office in the United States, and one of your patients has dual US/EU citizenship? Guess what? Your office is under GDPR. Whether it applies to you depends on your customer base, but you could have international customers and not even realize it.

Companies have handled this in three ways. One, they've ignored it, which I think is dangerous; two, they make their entire operation GDPR; or three, they separate out their databases for people who are in the EU and people who aren't, and then they have two different database systems and two different sets of records.

Three years ago, a bit before GDPR went into effect, I would tell all my clients whether they processed credit cards or not, they should be compliant with PCI DSS. PCI DSS is the payment card industry's digital security standard. There are twelve core things that you need to do to protect and secure your company online, and it's only required if you actually store and have access to credit cards. That's why many of small and e-commerce vendors will actually utilize third-party processing facilities where they don't touch, manage, or store the credit card themselves, so they don't have to be compliant with PCI. However, even if you don't process credit cards, it is still a good standard to follow to make sure you have the core areas of cybersecurity covered.

Data regulation is coming to the United States. It might be a year or two or three, and because the United States is so behind, when it actually gets passed, it's going to be rushed through the system. If you're starting from scratch, you will not have enough time to implement it, so your best bet is to start today. I recommend that you become GDPR compliant and use that as a foundational position. Even if you don't have customers who are EU citizens, it just sets you up for success first.

Protecting and securing client data is never a bad thing. It could also reduce your liability and lower your premiums if you do have cyber insurance in place. And most importantly, it's getting you battlefield ready. There are going to be regulations in the United States, so the sooner you adhere and the sooner you implement those standards, the better off you'll be.

CHAPTER 7 REVIEW

In this chapter, I've tried to raise your awareness on a national level. This is not about fear, uncertainty, and doubt, but we are in a cyber war. Countries are currently in our infrastructure and attacking us, but we are also attacking them. If information is factual and accurate, even if it scares somebody, people need to know about it. There are things that can take down our country, take down our infrastructure, and cause loss of life, and people have false or incorrect beliefs on how it can happen.

Watch what information you give to healthcare providers, and do extra verification when it comes to your finances. Realize that our electronic voting systems are not secure. Understand that there are guidelines for the data you store, especially for citizens of the European Union, but they may also affect your business, and the United States will, at some point soon, probably adopt its own. So it's best to begin that process.

We need to stop being afraid to spread the truth. If people don't understand the truth and know what the real problems are, they're not going to be able to fix, secure, and lock them down. The mindset we all need to start adopting should be to prevent what we can but to put a heavier emphasis on detection.

EIGHT

CYBERSPACE: A PLACE WITH NO BORDERS

We all live in parallel worlds. We spend a portion of our time in the physical world and a portion of our days in the virtual world, otherwise known as the internet or cyberspace. As the years pass, we are spending more and more of our life in cyberspace than we are in the real world.

I'm reminded of that firsthand as I'm writing this book; as I've said, we're in the midst of the COVID-19 pandemic and people are practicing social distancing. For the most part, many people are actually shutting themselves away from the physical world and spending ten or fifteen hours in cyberspace. People are running virtual meetings and virtual happy hours. What's interesting is I've had some friends point out that they are actually now able to "hang out" with people they normally wouldn't. In the physical world, you can only have happy hour with people in your town. Now that folks are distancing themselves from the physical world, you can have a virtual happy hour with somebody from Singapore, Australia, or Russia. Boundaries don't matter. In cyberspace, it doesn't matter where you are. You can interact in seconds with people who would otherwise take you twenty hours to travel to.

This is a mindset shift for many people because most of us have spent a good percent of our life in the physical world. Those rules and concepts are what drive many of our decisions. Some of us grew up without the internet, much less computers or cell phones. The only thing we knew was the physical world and, therefore, we learned that physical boundaries and physical separation is the way to protect and secure ourselves. If we can physically isolate and protect ourselves, that's the main level of security.

Now that we're shifting from living and operating in a physical world to a virtual world, all of a sudden those principles and rules that we learned are no longer applicable.

In the physical world that we live in, we can say that there are different countries, different states, different laws, and different regulations. And that's fairly easy because at any moment, we know where we physically are in the world. Right now, I am physically in Virginia, in the United States of America. I know that I have to follow the rules of the state of Virginia and the rules of the United States. Several months ago, when I visited Saudi Arabia, I knew that when I stepped off that airplane, I was now in the

Kingdom of Saudi Arabia and I had to follow their rules and regulations. Even though I'm a US citizen, if I did something that was legal in the United States but was illegal in Saudi Arabia, I could still get arrested and go to jail.

Now, with cyberspace, we have an interesting dilemma because all these boundaries are gone. We are one world and one country on the internet. On the internet, there is no concept of where you are. You can go anywhere, hit any country, get to just about any place in the world. In many cases, and this is the scary part, you are not even aware where you're traveling and accessing. Do you realize that if you're in Virginia and you're accessing a server in New York, that connection could go through Europe? Or Canada? Or even Russia? We don't really know which countries those packets are traversing and who could be monitoring them—and you. When we're at a site, we don't know where that site's located.

A lot of the websites of US companies that you're accessing may be in data centers or cloud providers in the Philippines, Venezuela, or other places around the world. So all of a sudden, when we are in cyberspace, it's a single entity. There aren't boundaries or customs. You don't have to show your passport. In cyberspace, international boundaries and laws disappear and you can access anyone, anywhere, anyplace. We truly live in one world, a single world, a world without boundaries.

Accessing a server in China, Russia, Canada, the UK, or the rest of Europe is just as easy and transparent as accessing a server in the United States. But here's where it gets tricky: Laws that dictate activity or behavior in physical countries still apply to the country the server is located in. If someone commits a crime in the United States, they are bound by US laws. But what if that person commits that crime over the internet and they're in a country where not only is that activity legal, but they don't have extradition treaties with the United States?

This is an interesting dilemma for the world we live in today. People are now committing crimes, and we know who they are and where they're located, but there's nothing we can do about it because the countries they're in don't allow them to be prosecuted or extradited to the United States.

The lack of understanding of life in the virtual world with the mindset of a physical world can create massive problems and issues.

For example, an individual who lived in Chicago was accessing a website in California. Let's just say that this website in California had adult content. The content was legal in California as well as Chicago. As far as this person was concerned, he was not breaking any laws. What he didn't realize was that the ISP (internet service provider) that he was using was actually going through Utah, and the content was illegal in the state of Utah. He is now being brought up on charges of transferring illegal content across state boundaries.

On one hand, he is breaking the law. This content is illegal, and he was accessing it across state boundaries. But on the other hand, he didn't know. If you talk to an attorney, they will always tell you that ignorance of the law is not a valid plea. Just because you didn't know you were doing something wrong doesn't mean you get off the hook. That's not how it works.

If ignorance were a valid plea, think of how many speeding tickets you could get out of by just saying to the police officer that you didn't know what the speed limit was. Even if you don't know the speed limit, it is your responsibility to notice it, and therefore you can still get a ticket for speeding. However, when I say that you can still get in trouble on the internet and be arrested for moving content across state boundaries, most people are not okay with that because they do not believe it's their responsibility to know where and when you're transferring and moving that content and the relevant laws. This creates a bit of a double standard.

That's just one example of the challenges that we have in front of us in terms of trying to protect and secure ourselves. In cyberspace, we don't necessarily know where we're going and what we're doing, but we're still bound by the physical rules and laws that underlie that activity. It's important as you go through your day that you're constantly aware:

Am I in the physical world or the cyber world?
Am I doing anything that could potentially get me in trouble?

I've also seen the increased use of encryption to protect and secure our data. Many people don't realize that many of the services and protocols you use don't necessarily encrypt the data by default. If you're now sending sensitive information in email, in a file transfer, or over the web, you have to recognize that it could go through countries that you might not want to see it.

If you're in Chicago and you're communicating with somebody in Boston, you would conceive that if you were driving in the real world, you would stay within the United States. There'd be absolutely no reason for you to drive through Canada or South America.

On the internet, the routing and the protocols don't necessarily work that way. It very reasonably could route you through China, Singapore, Canada, South America, or any other country, and any of those countries could potentially intercept or grab that communication. That's why it's important to encrypt your data and protect your information.

The good news is that most applications support encryption, and it's turned on in many cases. However, if you are going to use any application, it is important to spend five minutes reviewing the security settings and turning on the appropriate security that you need. Remember that cybersecurity is your responsibility, and this includes being aware of the applications and respective security they have available.

INTERNATIONAL HACKING, LAWS, AND YOUR RISK

Most of us now do our banking online. I still physically go to the bank, and I still write checks. I know there could be problems with that, but here's how I look at it: If I go into my bank in person to do an electronic fund transfer, I need to be present, provide my ID, and sign a piece of paper. Yes, people could get fake IDs or forge my signature, but, from the perspective of attackers, it greatly limits the playing field because now if somebody is going to try to steal money or do an electronic fund transfer in my place, they have to physically be in Virginia and walk into the bank. Increasing the activity that is required greatly reduces the number of people who will try. Could it happen? Yes. Is it less likely than in other scenarios? Absolutely.

Let's compare that to online banking. With online banking, all somebody needs is a user ID and a password, maybe an account number, but now they could be anywhere in the world. You've just taken the number of people who can break in and steal your money and increased it exponentially.

There are measures in place that try to protect you from this, such as two-factor authentication, but even in those cases, we have still seen attackers from around the world break in. As we described in Chapter 7, they will call you, posing as the bank, and ask you for a one-time security code that they've prompted the bank to send to you.

As we've discussed throughout this book, cybersecurity is about balancing risk with functionality, so I'm not saying you shouldn't or can't bank online. I'm just asking: Have you done the proper risk analysis?

Most people decide to use online banking because it's easier and more convenient. I'm advising you to step back and say, "Okay, this is easier and quicker for me, but it's also easier and quicker for the adversary." If they can now get into your account with less effort, is that exposure worth the benefit? If you say yes, it absolutely is, then you should continue doing that. If not, you might want to put some better controls and parameters in place. Recognizing the exposures and risk you face is one of the first and most critical steps.

Who would have thought that we'd live in a day and age where somebody in North Korea can steal your identity? They can compromise your information from your computer or numerous other locations.

That's really the bigger problem that you have to understand when it comes to the internet and cyberspace: Other people's bad decisions can put you at risk. I know that even though I don't do online banking, or give that information away, the bank still stores all my information online. If their systems or servers get compromised, my information can still be compromised. It's one of the reasons I emphasized earlier to use credit cards instead of debit cards. Once again, even if it's not you—it could be a clerk or vendor you gave your card number to, or a shop that might not be storing your information the way they should—it's potentially giving an attacker direct access to your bank account.

Today, cybercrime is big business. It's not some clever teenagers after school trying to make a buck. It is done by corporations. One of the biggest is RBN, the Russian Business Network. This is a company that has eight thousand employees, an office building, clear signage, salaries, benefits, and vacation time. They have a job just like you and me, except their job is to steal five thousand identities a day, and that's what they do. Just like RBN and the folks behind the WannaCry ransomware, which we mentioned earlier in the book, we knew who they were, where they were, and what they were doing, but because they were in Russia, and it's not technically illegal in Russia, and they don't have an extradition treaty with the United States, there wasn't much that we could do about it.

We couldn't even block the IP addresses of those responsible for WannaCry because, by now, the attackers knew our MO. Past attackers would use a single IP address, and then once we found it, whether it took days, hours, or minutes, and we blocked it, they would stop the attack, revamp, and come from a new IP address. WannaCry was savvy enough to switch IP addresses every five minutes. Essentially, they would start attacking, we would investigate, and we would find out the IP addresses, but it was almost useless to block them because they had already jumped to a new IP address.

These attackers understand the world we live in. They understand how defense and law enforcement operates, and unfortunately they're able to stay one step ahead. The part that frightens and frustrates me the most is that when you look at cybercrime, it is high payoff and almost no risk because even if you get caught, if you do it correctly, there's only a very limited chance, or even no chance, of prosecution. This is why cybercrime is on the rise.

If you're a criminal, I don't mean to insult you, and I thank you for reading my book, but most people who commit physical crimes are on the lower end of the IQ spectrum. Somebody who's very intelligent isn't typically going to rob a bank. In cybercrime, it's almost the opposite. Really smart people are the ones committing those crimes because they have the creative ideas, technical knowledge, and there's not a whole lot that's really stopping them.

This is highly, highly frustrating. It's why we need to work really hard and continue to push entities like the UN and others to build up these international laws. Knowing how things work between countries, we're probably still talking eight to ten years before that happens.

While many people don't think they can make a difference, you can contact your state and federal congressional representatives and tell them that this is a priority for you. The more people who do this, the more likely that your government officials will listen. Also, you can control who you vote for, and the more cybersecurity becomes a hot topic, like immigration and abortion, the more attention politicians will pay.

Cybersecurity is the next nuclear crisis. Politicians and the world need to come together to pass laws and treaties to protect us as one world. We need to lobby for this.

I pick on Russia a lot because they're very smart, active, and advanced in this space, so it's actually a compliment, not an insult, that I use them in my examples. I am going to do it again when I raise the hard question: Why would Russia want to cooperate when it comes to international law? In a lot of cases, the government is either partially or fully involved with these activities and they're making money off of it. So why would a lot of these countries want to participate in these different laws?

Getting to that point is not an easy route. In order to prevent a worldwide attack, it comes down to us detecting, stopping, and taking action. We cannot rely on law enforcement in this area just because the laws have not kept up with technology.

As I said in the previous chapter, traditionally, when we talked about cybercrime, we would usually determine the crime and tie it back to the country. So what I often say is if they're stealing intellectual property from Fortune 500 entities or the government, it's probably China. If the crimes are financially driven to promote monetary gain, it's probably the Russians. And if it's disruptive and taking down systems or servers, it's probably the Iranians or the North Koreans. However, even that is changing rapidly. We're seeing financially driven crimes coming from all those countries because it's easy, it's effective, and as we mentioned, it is zero risk.

We're never going to get to a point where we can completely stop cybercrime. Even if there are international laws, there are still criminals out there. The problem is that many companies and people are making it way too easy for attackers to break in. Many companies are not protecting their critical data. Many companies have systems and servers accessible from the internet that have critical data that is not encrypted nor protected. And many companies take more than three years before they actually detect a cyberattack against their business.

CHAPTER 8 REVIEW

The problem we have today is that the laws have not kept up with the technology that we're utilizing, even though on the internet, we're one world. You can go anywhere, any place, at any time. Somebody from Russia can go to the United States and vice versa without any passports, any immigration, any customs, any controls, any monitoring. It's easy. We've created an amazing technology infrastructure that allows worldwide communication, basically taking international boundaries and making them disappear, but we never thought about the legal ramifications of doing all this.

We don't have any international cyber laws. We don't have worldwide cooperation on this front. So now we have a scenario where somebody from any country, like Russia, can attack and break into the servers and steal money and information from citizens of the United States. Yet from a law enforcement perspective, there's not a lot we can do about it because we still have to deal with the physical laws, and we are pretty powerless in countries that don't support extradition or do not deem this activity to be illegal. Which is why cybersecurity must be your responsibility. In cyberspace, it's anarchy, and in anarchy, you need to protect yourself.

NINE

SURVIVING THE CYBER CRISIS

I'm often asked during television interviews if a "cyber 9/11" could really happen. My response? It's happening. We're in it.

We often think of a cyber 9/11 as a one-time devastating event, but there was a long buildup and it is happening now. Our systems, information, and data are and continue to be compromised. Over a billion records of US citizens including social security numbers and other critical data have been compromised over the years. That information can never again be protected. Many of us, including our children, are going to have to live the next few decades and beyond without having our social security number sensitive, secured, or private. It's an identifier we have for the rest of our lives that can never be changed, and it's out there. That shifts a lot of dynamics. We have to step back and recognize that we are living in an active war. The question is: What do we need to do to protect and survive these devastating events?

When I look back, I think about how the movies that were fictional are rapidly now becoming true to life. One of my favorite films, which I'm reminded of as I'm writing this section, is *Sneakers* with Robert Redford. The general premise is that somebody figures out a back door to all cryptography, and all sensitive personal information can be decoded, and there is no such thing as private data, secret data, or personal data anymore because anybody can read and access the information.

While we're not completely there yet, we are getting fairly close. As you know by now, most companies are not implementing encryption correctly. They store the keys with the data, and that really defeats the whole point when billions upon billions of records have been stolen and compromised. In almost every case, the data was encrypted—it's just the keys needed to decrypt the data were not properly protected and secured. We have this false illusion that there is security when, in reality, it is not set up correctly.

Another movie that comes to mind that's playing out is *WarGames* from 1983. In it, a teenager uses a war program that would dial random numbers to try to access different computers. He thought it was a video game, and then one number was answered, which turned out to provide access to a nuclear missile silo. He thought he was playing a game but was actually getting ready to launch a real missile against Russia and start a world war.

Watching it in 1983 might have made you roll your eyes that the US government could be so stupid as to have critical missiles and information directly connected to the internet, with guessable passwords and simple access. There's no possible way that someone could compromise a critical government facility without detection for days . . .

I look back now and I have to laugh a little, but I'll be honest with you, the laughter is pretty quickly turning to tears. One of the big trends I see as I sit here writing this book, as I've said, are interconnecting systems that were never meant to be interconnected. I will not go so far as to say that nuclear missiles are directly connected to the internet and can be accessed from your modem, but I will say that nuclear reactors, critical control, chemical processing facilities, and medical facilities that were never ever designed, built, or meant to be connected to the internet are starting to be interconnected without the proper security in place.

In *WarGames*, functionality comes first and cybersecurity is a distant second. Yet today, we haven't learned that lesson, and we still make that mistake, focusing more on the functionality and completely overlooking the security.

The other aspect of *WarGames* that hasn't changed is how many entities are still relying on simple, guessable passwords. There was a breach that happened less than two years ago with a large Fortune 500 company in which five hundred million records were compromised, and the user ID was "admin" and the password was also "admin." When I was starting my career as a professional hacker at the CIA in the eighties, if you told me that we would still have companies and entities with usernames and passwords of "admin" in the year 2020, I would have thought you were crazy.

We need to go back to the basics:

- Don't put critical data on internet-facing systems.
- Do not deploy outdated operating systems.
- With internet-facing systems, use only properly patched systems with strong passwords.

Over the last twenty-five years, we have failed to adhere to the basics and, unfortunately, we are setting ourselves up for being punished because we are getting sloppy. Not everyone is guilty of that, but there are entities out there that are still not taking cybersecurity seriously and this is creating a huge exposure point for all of us.

Another example of one of my favorite movies (that, to me, is now one of the scariest films of all time) is *Live Free or Die Hard*. You've just got to love Bruce Willis as John McClane. In it, there was a previous security professional who worked at the government and identified major flaws in their system, but nobody listened to him. He left and hired hackers to build various pieces of code to be able to break into these systems. Then, he held a "fire sale," where he could access all of the accounts and then delete everything, so now nobody had any record of how much money they had, of property they owned, of stock, or anything else.

In essence, if you added together all those accounts and transferred it to a single account before you did this, and put it in a safe offshore location, you could steal billions or trillions of dollars. Watching something like that in 2008 was fun because you'd think, at that time, we'd never have that much reliance on computers and the internet. Clearly, people would have paper records of that information. I watch that movie today and I'm not laughing, and I'm actually a little terrified because the question is: Could that happen now? Even on a smaller scale at one or two banks?

IT'S TIME TO GET SERIOUS ABOUT CYBERSECURITY

I'm not trying to instill fear. I'm trying to give you factual data and information. The fact is, we are more exposed than ever because we are totally and completely reliant on our computer systems, and if those computer systems went down or weren't accessible or were lost, most people don't have backup paper records.

What if somebody showed up at your front door with the sheriff and said, "This is my house."

You reply, "What do you mean? This is *my* house."

They say, "No, here are the deeds. Here are the documents filed with the county, and this is my house. Do you have paper copies? Can you prove it?"

So you say, "Let me call up the county."

When you do, they say, "Oh no, the house belongs to John Smith, not Eric Cole."

What would it really take to do that? How many databases or systems would you need to compromise? How many records would you have to alter in order to falsify ownership or hide or deny somebody who really owned that piece of property?

When I give a keynote, it usually starts off with a lot of nay-sayers and rolling of the eyes, as you yourself might have done with this example. But when people really start to see the lack of protections or verification that most of them have on their bank accounts, investments, and homes, it starts to get really scary very quickly.

It's not about living in fear but recognizing the reality of this occurring and making sure you're properly prepared and protected.

One thing I and all security professionals can say is that a large-scale cyberattack is coming. It is inevitable. Look at the data over the years. The amount of cybercrime, money stolen via cybercrime, digital information stored, and the number of vulnerabilities on systems are all increasing year after year. If you just do the basic math and you run any projective curves that you want, they all render in the same conclusion: Cyberattackers and big business countries are actively partaking in cyber war, and it is going to continue to increase for the foreseeable future because there are no international laws or regulations to be able to monitor or control these types of attacks.

The question is: Can you protect yourself, and do you have a basic level of cybersecurity to back up and control your information?

When we talk about someone breaking in and taking down entire banking systems, modifying all the medical records for an entire country, or taking over the water processing or the utility companies, most people, even security people, can't comprehend it because it hasn't been done—yet. Unfortunately, the vulnerabilities and exposures exist to make it possible, along with the rise in cyberattacks. And if we don't start fixing it, raising awareness, and working together, it will happen, and it is currently happening more often than most people are aware of.

WHAT CYBER 9/11 MIGHT LOOK LIKE

When most people think of some massive cyberattack, they usually think of traditional infrastructure, like cyberattacks against air traffic control systems, power grids, utilities, communication, and government systems. Those are definitely feasible. To be fair, there is redundancy and recovery built into those systems. Still, if you remember eight or ten years ago, a portion of the United States lost power and the government claimed it was user error, and I've seen a similar thing happen in many states. But was it user error or was it a cyberattack?

If you say it was a cyberattack or cyber compromise, people would freak out. We're not ready to accept this as reality because it's incredibly scary. But even if technically it was a user error, that means there are exposures. If one person could mistype a key and take down the entire power grid for a section, what makes us think that an attacker would not be able to do the same?

I personally think the attacks are going to be a little less traditional and much more data focused. What controls the world today is not electricity, water, and airplanes. What controls the world today is data. The person with the most data wins. We've seen that with a bunch of companies.

Let's look at Google, for example. Everyone thought Google was a search engine, but they knew a long time ago it was all about the data. They started setting up Gmail and Google Docs and it's all free, free, free. No, it's not. It's data, data, data. You give me all of your data because I give you an online document that you can share with others. I'll take that deal every day of the week because your data is worth a heck of a lot more money than that application you're utilizing. As I said before, wouldn't it make a lot more sense for you to just pay $79 to use the app and have your data secure and protected? It would be a much better deal for you, but not a better deal for them, and they know that.

We've already seen ransomware attacks within a short period of time steal over $20 million, and these are one-off cybercriminals doing it. What if it was more coordinated? What if there was a full-blown coordinated ransomware attack across multiple industries, businesses, and organizations, all at the exact same time? Now all of the financial, healthcare, and tax data is held ransom, and none of that information is accessible or available. Now what?

We've seen hospitals and banks under attack and compromised, but what if they all happened together? As I mentioned, as I write this, six months into COVID-19, some hospitals are overflowing and there's not enough equipment. What if somebody now launched a cyberattack against those hospitals? How much damage and loss of life would occur? On the other hand, how much would somebody be willing to pay to recover that data?

We've also seen attacks against stocks, stock market devaluation, and short selling. What if we started seeing that more persistently? What if every day, somebody went in and was skimming or short selling $20 million? Multiply that by five hundred days and you can start to see the impact.

Probably the one that scares me the most is when you start getting into cryptocurrency and Bitcoin. You essentially have a fabricated currency that has no backing and no value except people believe that it has some value. What if we continue to create these currencies that really have no backing other than perceived value, and then all of a sudden they crash, they go down, they disappear?

WHAT YOU CAN DO

As a citizen of a country, you can and should start to put pressure on lawmakers. In the United States, you can talk to your congressional representatives and make them aware that cybersecurity needs to be a priority. You can start voting for entities or individuals that understand and promote proper cyber laws and cyber regulations. Because whether we like it or not, the United States is way behind.

When it comes to you, your business, and your family, you can begin to ask better questions and start to protect and secure your information. One of the best things you can do for your critical documents and critical information is to start printing out some hard copies as backup. I still have hard copies of my tax records for the last several years. I periodically print out my bank records. I have hard copies in multiple spots of my land deed and the mortgage papers for my house. Having access to this information, while it might seem simple or trivial right now, could be huge if something happened.

Start protecting and backing up your data, whether it's for your business or your home. That data should be in multiple spots. You should have data on USB drives that are not connected to your computer, maybe locked in safes, or even at third-party locations, like safety deposit boxes. These are not hard things to do, but they could make all the difference in terms of being able to protect and secure your information.

And don't forget the little and, yes, maybe more inconvenient things, like setting up stronger passwords, having two-factor authentication on all your accounts, and setting up alerts so that if anyone tries to access your business accounts, your personal accounts, or other information, you get alerts or notifications.

CHAPTER 9 REVIEW

Our systems, information, and data are and continue to be compromised. Cybersecurity doesn't mean you have to live in fear; it's about recognizing the reality of this occurring and making sure you're properly prepared and protected.

A large-scale cyberattack is coming. It could be in our infrastructure, but most likely it'll be stealing massive amounts of our data and either holding it ransom or using it to cause large-scale chaos in other ways. While you can't secure the world, you can do a lot to protect your piece of it. And if you take away only one thing from this book, remember that you and your business are a target, and cybersecurity is your responsibility.

Conclusion

Ten Lessons to Remember

I've covered a lot in this book, so I want to close out by providing you with a summary of the top ten lessons to remember to help you ensure your cybersecurity:

1. You are a target.
2. Cybersecurity is your responsibility.
3. Security is typically built in but not turned on by default.
4. Email attachments and web links are the most likely danger points to cause harm.
5. Always focus on critical data and critical information.
6. Ask questions to understand the risk and exposure.

7. Always back up your critical data to offline storage.
8. In cyberspace, there is no delete button.
9. Detection via monitoring is the key to security.
10. Always act under the premise that you are compromised.

Let's look at each of these in more detail.

1. You are a target.

One of the most important mental shifts that you need to under-take in order to protect yourself, your business, and your family in cyberspace is to know you are a target, no matter what. As long as you walk around thinking that nobody wants your personal information or business data, that's going to paint an even brighter target on your back because you're going to be too relaxed and maybe even get sloppy. If you don't understand that you're a target, you're not going to put basic measures in place and practice good security. It's like practicing good hygiene. If you believe you are never going to get sick, you might not wash your hands or use caution, and as soon as you become careless, that increases your probability of getting ill. In cybersecurity, if you believe that you're not a target and you don't have a sense of vigilance or put strong measures in place, that will increase the probability that you get compromised.

I've been doing this for thirty years and trust me: It doesn't matter who you are, what you do, or where you live. I could fill volumes of books on actual attacks and breaches where people just like you have been targeted and compromised. They all say the same thing: I wish I had listened to you when you told me I was a target, but I didn't believe you. There were companies that went out of business and countless identities stolen because they didn't heed my warnings.

Learn from other people's mistakes and recognize that over the course of your lifetime, and probably more likely over the next twelve months, you, your family, and your business are going to be compromised. The more aware you are, the more protective measures you put in place, the better off you will be overall.

A key is to implement strong passwords and two-factor authentication. It is critical that you turn on two-factor authentication whenever possible, and if it's not an option, you always want to have strong passwords and different passwords for each account. You can absolutely use a password vault program but still make sure you have different, good, strong passwords for each account and change them periodically. That way, if for some reason you do get compromised, the impact to you is minimal.

Most importantly, never, ever give out your authentication information. The number of social engineering attacks where people are trying to manipulate you to get information is increasing tremendously. Adversaries are advancing. Instead of trying to use super, robust exploits, it's often easier to pretend that they're a legitimate entity on the phone and just ask you for the authentication information. Believe it or not, it is one of the simplest but most effective methods of compromise.

Every single day, multiple times a day, remember that you and your business are a target in cyberspace and constantly ask yourself what you can do to minimize and reduce that risk.

2. Cybersecurity is your responsibility.

I often meet with executives who run million- and billion-dollar businesses, and they point to an entire department whose only responsibility is cybersecurity. They'll ask me why is it important for them personally to know about security when they have a staff to do that. We can give you a very secure environment and computer, but if you are allowed to have the functionality that you need to do your job, there is still exposure to make mistakes and for your cybersecurity to be compromised. So, even if you're a CEO, you need to take personal responsibility.

Every time you add functionality to your system, you are reducing security. In order to give you a functionally rich computer and network that allows you to do your job, it will include some security risks and exposure.

Be careful of what you click on and open on your system. As you're working on your computer, it's important that any time you're about to perform some external action, whether it's visiting a site, clicking on embedded links, downloading content, or opening an attachment, always ask yourself: What are the risks and exposure? This book gives you some ideas of what you can do to be secure, but you have to adopt the mindset that cybersecurity is your responsibility.

3. Security is typically built in but not turned on by default.

The good news is we are at a point right now where companies are recognizing that security is important, and security measures need to be put into place. The bad news is they do not believe that consumers are willing to tolerate all of the inconvenience that comes with that security. Instead of turning it on automatically, most vendors provide the security, but you and/or your team must take the time to locate and activate those tools. I think it would be better to provide a locked-down version instead of having to spend twenty or thirty minutes turning on the security features that are in the app but not turned on by default. Some apps, perhaps, might not work right away because of the added security, but then at least you would be aware and able to turn off the security if you chose.

As mentioned, the COVID-19 pandemic has brought an interesting inflection point to security concerns. Many people have started to use Zoom for videoconferencing. Like most companies, Zoom took the approach of building in security, but not turning it on by default. They got so much negative publicity about unauthorized people joining calls (called "Zoom bombing") that they actually came out with a new release that had all the security automatically turned on. It came with some inconveniences, but based on all of the exposures, it seems that the public now expects it. I hope that this becomes the new model, and by the time you read this book, companies turn on security features automatically. But I do not believe all companies will, so please remember that it's your responsibility to recognize, enable, and correctly configure the security features of any device, program, or application you or your team use.

4. Email attachments and web links are the most likely danger points to cause harm.

It's important to remember that any application could potentially be an exposure point and create weaknesses in your environment, but if I had to pick two of the most dangerous applications on the planet, they would be email clients and web browsers. Specifically, the exposure is in email attachments and malicious links that users click on.

Be aware of this and develop good habits. Never click on an attachment in an email unless you are 100 percent sure it's legitimate. Even better, just make a general rule that you will not open up any email attachments. I know that sounds insane, but email was never meant as a file transfer mechanism. It was meant just for basic communication. If you have entities that you need to share documents or files with, it's much better to set up a secure file-sharing system or Dropbox-like system to which authorized people have access so it can be monitored. File transfer systems can also be targeted, but it's not as common or as easy for the attacker. If you have strong passwords and proper monitoring, it tends to be a more secure option than allowing attachments in emails.

If you or your organization prefers to stick with email attachments instead of a file transfer site, another option is to use a non-Windows-based system, such as an iPhone, iPad, or Android to do an initial pass, since most attacks today are on Windows devices. If you do make a mistake and accept a malicious attachment, there will be less exposure to you and your company on those other operating systems.

If you can't get an iPad or similar device to check email, another option is to utilize two computers: reserve one computer to contain your critical data and use the other computer only for email and accessing the internet. If the computer connected to the internet gets compromised, the impact is minimal because there's no critical data on that system. For instance, you might delegate your desktop computer as the secure one, and keep all your critical data on that, and use a laptop to check email and cruise the web. If you do use the laptop for work, send those files to your desktop at the end of the workday, and remove anything critical from the laptop.

When it comes to web links, it's important to remember that when you go to a website, even if it's a legitimate site, many have ads that come from third parties. They have absolutely nothing to do with the actual entity you're visiting, nor did the website validate or approve them. If something looks unusual or too good to be true, don't click on the link. I'm currently seeing so many people getting conned out of money and having their identities stolen because they are going to news sites to read about COVID-19, and ads are popping up for hand sanitizer, masks, and other mechanisms to be safe and secure. These items are in such high demand, and people are so desperate to protect themselves and their family, they click, and their information gets stolen. In one particular case, an ad said they were selling sanitizer, but because of such high demand, their credit card processing system was down and they were only accepting debit cards, with only fifteen items left in stock, which they estimated would be gone in ten minutes. They were pushing for an emotional response. And debit cards are very dangerous because the money comes directly out of your account, and that's exactly what the attackers were targeting. These fraudulent types of ads are getting very popular, so do the research and be careful, and remember: If a deal is too good to be true, it probably is.

Make sure you are very cautious and alert and only click on legitimate links. My approach for any website that you visit on a regular basis is to download the app for the site, if they have one. This works well with Amazon and banking sites. Or you should have a legitimate site bookmarked and always go through the application that bookmarked it.

5. Always focus on critical data and critical information.

The goal of cybersecurity is not to prevent all attacks but to minimize or control the damage. The difference between a major attack and a minor attack is the data that is compromised. If you want to take a proper approach to cybersecurity, it should focus on your critical data on a regular basis for your business and even for your personal life. You should always ask yourself: What is my critical data, where is it located, and who has access to it? The more that you get up to speed with protecting, securing, and locking down your critical data, the better off you will be with cybersecurity.

This is always an interesting balance since, on one hand, you don't want your data all over the place because then it's hard to control and manage and very easy for the adversary to get access to it. On the other hand, you don't want to have your data only in one spot because if it's compromised, you could have a major availability problem if you are hit with ransomware or if the system crashes. There's a balance where you want to have your data in more than one but less than any unnecessary locations. Also, having some cloud-based backups can help out. I recommend that you store your data in a few different locations where you can still control and manage it and make sure it's properly protected, secured, and locked down.

6. Ask questions to understand the risk and exposure.

I don't expect you to be super technical. I don't expect you to know how to configure firewalls or understand the different types of encryption. I do expect you at a strategic level to be able to ask better questions. Remember: "Smart people know the right answers. Brilliant people ask the right questions." With cybersecurity, it's important that you don't make assumptions and you ask better questions.

When one of the major hotel chains had five hundred million records compromised, if the executives and management would've asked better questions, those questions might have been: Do we have any systems visible from the internet that still contain critical data? Do we have any systems that are accessible from the internet or not patched and up to date? Did anyone know those legacy systems were still active, not patched, and contained data that was not properly protected? In this situation, there was a group of people who knew about these exposures and vulnerabilities, but the executives were unaware.

That disconnect between the two creates problems and major security breaches. In every security breach that I've worked on, the one commonality is that the executives are unaware of the real exposure, and somebody on the technical team knew about the exposure and was frustrated that nobody was listening to them about it. It's this disconnect between the high-level strategy and the technical people that creates these exposures.

One of the best ways of addressing and then fixing this is through increased communication. When you have two entities that speak different languages, the way you increase communication is by asking better questions. Throughout this book, I gave you clearly marked questions that you can ask. By continually asking questions as a business leader and executive, you can avoid many breaches and open up the communication channel.

7. Always back up your critical data to offline storage.

We talked about knowing where your critical data is and limiting its exposure, but as ransomware attacks continue to increase, it's important to not only have replication of data through data centers and cloud providers but also to have offline storage and backups. That way, even if you do get hit with ransomware or your data gets compromised, you can still recover the information. You should have backups and hard copies of personal information, too, like your tax records, your bank account information and statements, and the deed to your house. Otherwise, if all of that is only online in one location and it gets compromised, you have no proof of ownership or validation of paying taxes. Paper copies of critical information might seem old-school, but it might save you. And nothing beats having an external hard drive that you back up your data to once a week or once a month (depending on how much it changes) and just putting that in a safe, secure location.

8. In cyberspace, there is no delete button.

Information will always exist forever, even if you believe you deleted it. One of the best pieces of advice I can give you is that before you hit send, save, or post, always ask yourself if you want that information to live forever. I see politicians, executives, and even friends post inappropriate things to social media and recognize the errors of their ways and remove it within fifteen minutes. Years later, I can find and locate that information in cyberspace because it still exists. While you might think that you can locally delete a copy off your device or that social media is not storing that picture you took, there is so much replication that the information is still out there. Emails, texts, or messages via any other transmission are fair game, too, and will always have some replication, and that information will always be available. Before you snap that inappropriate picture of yourself, remember there is no delete button on the internet and that information will last forever.

9. Detection via monitoring is the key to security.

One of the phrases that I'm probably most known for creating is "prevention is ideal but detection is a must." You cannot prevent all attacks. The only way to do that is by having no functionality. As long as you allow some functionality, there are going to be some exposure points. The key to security is proper detection. Turn on monitoring for all critical accounts. For example, my company utilizes cloud-based services, but for all of our critical accounts, we have monitoring in place so if anyone tries to log in from a new location, we are notified whether they're successful or denied access. It helps us to know what's going on with the business. I also have monitoring on for my personal bank accounts, so I am notified if anyone tries to access or transfer money or do anything that could have any financial impact. This is one of those areas where receiving notifications could be a little inconvenient, but the exposure of a major breach, because you didn't monitor access, is far worse than a little extra work. Monitoring is one of the best practices to protect your business, life, and your family.

10. Always act under the premise that you are compromised.

Perhaps the point I would like to drive home the most and leave you with: If you always assume that you are compromised and the adversary is already in your network, then you are always going to be careful and look for the adversary. It's going to keep you on guard and properly focused. It brings us back to the beginning where most people don't think they're a target, so they let down their guard. It comes full circle to recognize that you are probably already compromised, and if you're not seeing the signs of compromise, it's not because it didn't happen, but because you're not looking in the right place.

———

The threats in cyberspace are real, and it is critical that you take preventive action to protect yourself, your family, and your organization. While breaches are going to happen, it is critical to detect and respond in a timely manner. May all of your breaches be minor, and Godspeed in keeping your assets protected.

Index

A

access
  cloud services, 114–115, 125
  to critical data from internet, 43–45,
    74–75, 82, 110, 151, 211, 216
  to information. *See* availability of
    information
ad links, 63, 229
air-gap networks, 165–168, 177–179, 215
air traffic control systems, 162–163, 165,
  219
alerts. *See* notifications
Alexa, 21–23
alteration of information, 31, 36, 167,
  174–175, 217. *See also* integrity of
  information
Amazon Web Services (AWS), 106, 132
Android devices, 78, 91, 97–98, 101, 228
antivirus programs, 38, 40, 72
App Store, 91, 100
Apple (Mac) computers, 64, 66, 78, 178
Apple Technologies, 128
applications (apps), 89–92, 98, 100–103,
  206, 220
ARPANET, 25
asset inventory, 82, 152, 178
authentication, 28, 56–59, 65–66, 98, 100,
  103, 114–115, 125, 191, 207, 222, 225.
  See also two-factor authentication
  (2FA)
availability of information
  attacks against, 167, 231
  cloud services, 124
  healthcare sector, 174–175, 182
  and ransomware attacks, 32–33, 36,
    231. *See also* ransomware attacks
  as tenet of cybersecurity, 30–32, 36,
    124, 143, 146, 174, 182

B

backups, 19, 32, 116, 182–183, 197, 217,
  221–222, 224, 231–233
bank accounts, 27–28, 31, 59–60, 185,
  233–234. *See also* credit cards; debit
  cards
Bezos, Jeff, 40, 93–94
Bitcoin, 21, 28, 221
blackmail, 62–63, 96–97
breaches of cybersecurity. *See*
  cyberattacks
burner phones, 49. *See also* mobile
  devices

C

caller ID spoofing, 190–191
Canada, 158, 170
cell phones. *See* mobile devices
Central Intelligence Agency (CIA), 6–8,
  38, 45–46, 89, 147, 215
chief information security officers
  (CISOs), 42, 137, 144
China
  apps developed in, 100–101
  attribution of attacks, 171–173
  electronics made in, 26, 159–161
  email breach, example of, 28–29
  infrastructure in US, targeting, 26,
    158–161, 169–171, 173
  intellectual property theft, 139–140,
    159, 169–170, 211
  and international law, 133–134
  servers located in, 203
  small businesses, attacks on, 15
  travel to, security precautions,
    49–50, 66
CIA (Central Intelligence Agency), 6–8,
  38, 45–46, 89, 147, 215

CIA (confidentiality, integrity, and availability), tenets of cybersecurity,
30–32, 36, 124, 143, 146, 174, 182
Cisco, 132, 161
CISOs (chief information security officers), 42, 137, 144
cloud services, 32, 105–125, 132, 231
communication
email. *See* email
between executives and security
team, 137, 232
online versus face-to-face, 39–40,
95–96
verifying, 189–190
compromise, assumption of, 14–15, 70,
140, 148, 224, 234–235
computers. *See also* operating systems
(OS)
Apple (Mac), 64, 66, 78, 178
laptops, 26, 49–50, 64–66, 78, 151,
159–161, 176, 229
multiple computers, use of, 64–65,
78, 229
confidentiality of information
breaches, 139
cloud services, 124
Health Insurance Portability and
Accountability Act (HIPAA),
174–175, 182
as tenet of cybersecurity, 30–31, 36,
124, 143, 146, 182
configuration management, 82, 109,
152
COVID-19, 57, 77–78, 132, 136, 201, 220
credit cards, 13, 47, 75, 185–188, 191,
197–198, 208
credit freeze, 191
criminal activity, 174–175, 185, 204–205,
208–211, 218. *See also* international
laws
critical assets, 134–138, 153–155
critical data
accessibility of from internet, 43–45,
73–75, 82, 110, 151, 211, 216
app permissions, 91–92
asking questions about, 74, 81–82, 85,
99, 118–119, 145–146, 150–151, 153,
231–232
backups. *See* backups
categories of, 151
and cloud services, 110, 118–119
encryption. *See* encryption
exposure points, 82, 140–150, 155, 216

focus on, 223, 230
identifying, 145–146, 151, 155, 230
intellectual property. *See* intellectual property
location of, 115–119, 145–146
on mobile devices, 98–100
monitoring for attacks, 80–81. *See
also* monitoring
multiple computers, use of, 64–65,
78, 229
personally identifiable information
(PII), 3, 123, 130, 151
prioritizing, 146, 155
questions to ask, 74, 81–82, 85, 99,
118–119, 145–146, 150–151, 153,
231–232
security analysis, 143, 146, 153
security measures, 80–83, 85
value of, 2–3
critical infrastructure. *See* infrastructure, cyberattacks on
cryptocurrency, 21, 28, 221
cryptographic key, 75–76, 82–83, 85,
119–121, 214. *See also* encryption
cryptography. *See* encryption
cyber 9/11, 213–214, 218–222
cyber cold war, 24, 162, 172. *See also* infrastructure, cyberattacks on
cyber currency, 21, 28, 221
cyber insurance, 198
cyber murder, 174–175
cyber war, 25, 157–158, 198, 218
cyberattacks
asking questions about, 81, 85
assumption of compromise, 14–15,
70, 140, 148, 224, 234–235
attribution of, 171–173
cyber 9/11, 213–214, 218–222
frequency of, 1–2, 51–52
inevitability of, 51–55
on infrastructure. *See* infrastructure,
cyberattacks on
large-scale, 218
long-term effects of, 3
media coverage of. *See* media
power grid, 23, 67–68, 173, 219
proactive response to, 15, 51–52, 70
reporting, 69–70, 85, 130, 154, 164
scope of, 2–3
targets of. *See* targets of cybersecurity attacks
undetected, 3, 53–55, 67–68, 79–80,
140, 211

cybercrime, 174–175, 185, 204–205, 208–211, 218
cybersecurity department. *See* security department
cyberspace
  international laws, need for. *See* international laws
  and physical boundaries, 39, 133–134, 201–206, 211–212
  rules for survival in, 12
  time spent in, 37–40, 65, 201–202

D
dark web, 12–17, 50, 169
data centers, 41, 106–108, 112–113, 115–118, 203, 232
data, critical. *See* critical data
data theft, 3–4, 13–15, 42, 47, 68, 75, 185
data, value of, 2–3, 13
debit cards, 43, 185–186, 208, 230
detection. *See also* monitoring
  asking questions about, 80, 85, 146
  of hacking, 71–73, 80–81, 85
  infrastructure attacks, 164, 199. *See also* infrastructure, cyberattacks on
  versus prevention, 44, 163–164, 199, 234
  slow performance of servers, 54–55, 81
  undetected attacks, 3, 53–55, 67–68, 79–80, 140, 211
disaster recovery, 116–117
disclosures. *See* confidentiality of information; reporting breaches
double authentication. *See* two-factor authentication (2FA)
Dropbox, 122, 228

E
elections and e-voting, 79, 192–196, 199
email
  attachments, 30, 32, 64, 71–73, 77, 82–85, 96, 142, 152–153, 223, 226, 228–229
  compromised, 58–59
  detection of hacking, 71–73
  embedded links. *See* embedded links
  falsified, 27
  ILOVEYOU virus, 34–35, 147
  phishing schemes, 29, 39, 58–59, 61–62, 66, 77–78

  separate computer for, 64, 78
embedded links
  ad links, 63, 229
  in email, 30, 32–33, 63–64, 73, 77, 82–85, 152–153, 183, 223, 226, 228–229. *See also* phishing
  text messages, 96. *See also* text messages
encryption
  applications, 100, 103
  asking questions about, 82, 119–120, 122
  cloud services, 119–122
  critical data, 74–76, 82, 85
  cryptographic key, 75–76, 82–83, 85, 119–121, 214
  cryptography, 119–120
  importance of, 205–206
Endpoint Antivirus, 40
Equifax breach, 3, 13, 191
European Union, 196–197, 199
exposure points, 82, 124, 140–153, 155, 216, 223, 228, 234
extradition treaties, 133, 204, 208, 212

F
Facebook, 46, 60
false alarms, 55–56, 72
FBI, 148
file transfers, 65, 205, 228. *See also* Dropbox
finance sector
  account notifications, 186–188, 191
  bank accounts. *See* bank accounts
  caller ID spoofing, 190–191
  credit cards, 187–188
  credit freezes, 191
  and cybercrime, 185
  debit cards, 185–187
  large-scale cyberattack on, 218–220
  North Korea, attacks from, 185
  notifications, 59–60, 186–188, 191, 222, 234
  phishing emails, 188–190
  regulations, 123
  Russia, attacks from, 185. *See also* Russia
  two-factor authentication, 191
  verification out-of-band, 189–190, 199
financial impacts, asking questions about, 84, 144
Financial Monetization Act, 123
firewalls, 71, 110, 149, 231

fraud
    ad links, 229–230
    credit card fraud, 13, 186–187. *See also*
        credit cards
    financial fraud, timeframe for
        reversing, 27–28, 58–59, 189
    fraudulent transfers, 185–186, 188–190
functionality versus security
    air-gap networks, 166–168. *See also*
        air-gap networks
    analysis, 17–23, 35–36, 42–44, 65,
        144–145, 207, 226
    apps, 100, 144
    asking questions about, 20–23, 36, 44
    cloud services, 105, 108–110
    elections, 192
    online banking, 207
    remote workforce, 57

G
General Data Protection Regulation
    (GDPR), 196–198
global nature of cybersecurity threats,
    23–26. *See also* specific foreign
    countries
Gmail, 220
Google, 45–46, 65, 219–220
Google Docs, 122, 220
Grumman Aerospace (Northrop Grum-
    man), 6–7, 37

H
hackers and hacking
    critical assets as target, 134–138, 155
    data theft, 3–4, 13–15, 42, 47, 68, 75, 185
    detection, 71–73, 80–81, 85
    disclosing attacks, 69–70
    discovery of, downplaying, 69–70,
        78–80, 85
    exposure points, 140–152, 155
    in foreign countries, 133–134
    international laws, 209–212
    potential harm to business, 138–140
    prevention, 70–71, 80–85
    targets of, 73–78, 84–85. *See also* tar-
        gets of cybersecurity attacks
    undetected attacks, 67–68, 80
Hackers Beware (Cole), 73
hard copies. *See* paper records
hardening, 115, 125
Health Insurance Portability and
    Accountability Act (HIPAA), 123,
    174–175, 182

healthcare sector
    air-gapped devices, 177–179. *See also*
        air-gap networks
    asking questions about healthcare
        data, 124
    cloud services, 123–124
    confidentiality, integrity, and avail-
        ability issues, 182–184
    cyber murder, 174–175
    cyberattacks on, 173–184, 218–220
    equipment connected to internet,
        177, 182
    functionality versus security, 182.
        See also functionality versus
        security
    Health Insurance Portability and
        Accountability Act (HIPAA), 123,
        174–175, 182
    network connections, 176–182
    operating systems, outdated, 176–177
    personal information, providing,
        183–184, 199
    security issues, 178–181
    WannaCry attack, 32, 34, 182–183,
        208–209
home offices, 11–12, 33, 56–57, 132
home ownership, proof of, 217
host discovery, 178. *See also* asset
    inventory

I
identity theft, 3–4, 13, 75, 185, 208
ILOVEYOU virus, 34–35, 147
Industrial Control Systems (ICS), 54,
    165–168. *See also* infrastructure,
    cyberattacks on
information warfare, 159. *See also* cyber
    war
Infosecurity Hall of Fame, 9
Infrastructure as a Service (IAAS),
    112–114, 121, 124
infrastructure, cyberattacks on
    air-gap networks, 165–168, 177–179,
        215
    air traffic control systems, 162–163,
        165, 219
    attacks on other countries by US,
        158–159
    from China, 26, 158–161, 169–171, 173
    critical infrastructure, 165–173
    detection versus prevention, 44,
        163–164, 199, 234
    extent of, 157–159, 162–163, 198–199

Industrial Control Systems (ICS), 54, 165–168
internet, connection to, 166–168
large-scale attack, 218–219, 222
media coverage, 164–165
from North Korea, 158–159, 170–171
nuclear reactors, 24, 54–55, 163, 165–167, 177, 194, 215
prevalence of, 198–199
regulations, compliance with, 196–199
retaliation, 171–173
from Russia, 158–159, 162, 172
security, 165–169
source of, determining, 169–171, 173
Stuxnet, 54–55, 162–163
Supervisory Control and Data Acquisition (SCADA), 165
vulnerability scans, 167–168
Instagram, 46
insurance, 198
integrity of information, 30–31, 36, 124, 143, 146, 174, 182
Intel, 23
intellectual property
attacks on from China, 139–140, 159, 169–170, 211
as critical data, 42, 151
as target, 42, 129–130, 135, 144, 157, 159
international laws, 133–134, 196–198, 208–212
International Traffic in Arms Regulations (ITAR), 116
internet
connectivity points, 25–26, 171–172
dark web, 12–17, 50, 169
disconnecting from, 25, 42, 172–173
inbound connections, 80, 85, 171–172
origins of, 25
outbound connections, 80, 85, 171–172
physical boundaries, lack of, 203–206, 211
and remote workforce, 57
internet protocol version four (IPv$_4$), 171
internet protocol version six (IPv6), 171
IP address, 106, 133–134, 169–171, 209
iPads, 64, 78, 163, 228–229. See also mobile devices
iPhones, 64, 78, 97–98, 101–102, 163, 228. See also mobile devices
Iran, 15, 54–55, 162–163, 169–170, 172, 211

IT department, 54–55, 66, 81, 108–109, 139, 142
ITAR (International Traffic in Arms Regulations), 116

L
laptops, 26, 49–50, 64–66, 78, 151, 159–161, 176, 229
legacy systems, 232. See also patches
legislation
and attacks from foreign countries, 133–134, 202–212, 218, 221
disclosure laws, 69–70
extradition treaties, 133, 204, 208, 212
Health Insurance Portability and Accountability Act (HIPAA), 123, 174–175, 182
need for, 221
privacy laws, 196–198
links, 229–230. See also embedded links
Linux, 64, 178
Live Free or Die Hard (film), 216–217
Lockheed Martin, 9, 15–16, 130–131, 137
log files, 76

M
Mac computers, 64, 66, 78, 178
"made in China," 26, 159–161
malware, 26, 55, 64, 66, 72, 90, 96–97, 160–163
Marriott breach, 2, 13, 24, 74
McAfee, 9
media
author's role as cybersecurity expert, 1, 9
coverage of breaches and attacks, 1–2, 30, 35, 75, 164–165
metrics, 81, 85, 111, 115, 153
Microsoft, 17–18, 74, 132
mobile devices
apps, 89–92, 98–103
cell phones, 17, 26, 40–41, 49, 59, 64, 78, 87–93, 96–103, 159, 163, 228
cyber blackmail, 62–63, 96–97
cyberbullying, 95–96
deleted information, 92–95, 102
encryption, 100, 103
hackable, 87–89, 102
iPads, 64, 78, 163, 228–229
location services, 88–89, 101
malware, 96–97
personal information on, 88
photos, 92–94

mobile devices (*continued*)
  privacy settings, 101–102
  security, 97–98, 103
  spamming, 96
monitoring
  asking questions about, 81
  attacks from foreign countries, 170,
    218
  cloud services, access to, 107, 111,
    114–115, 124–125
  dark web, 13–15
  of email, 188. *See also* phishing
  embedded in technology compo-
    nents, 160–161
  false alarms, 55–56, 72
  file transfers, 228
  importance of, 224
  inbound traffic, 80, 85
  infrastructure attacks, 158, 160
  intrusion detection, 70–72, 110
  as key to security, 224, 234
  and location of servers, 203
  outbound traffic, 80, 85
  social media accounts, 48, 65
multitiered architecture, 44

N
nation-state attacks, 15. *See also* specific
    foreign countries
National Health Service, 32, 34, 182,
    208–209
North Korea, 15, 25–26, 133, 158–159,
    169–171, 211
Northrop Grumman (Grumman Aero-
    space), 6–7, 37
notifications
  false alarms, 55–56, 72
  financial transactions, 58–60, 66,
    186–188, 191, 222, 234
nuclear reactors, 24, 54–55, 162–163,
    165–167, 177–178, 194, 215
Nuclear Regulatory Commission
    (NRC), 194

O
Obama, Barack, 9, 88
Office of Personnel Management
    (OPM), 74–75
offline storage, backups to. *See* backups
onion routers, 12
online activity
  banking, 206–208

cyber blackmail, 62–63, 96–97
cyberbullying, 95–96
  increase in, 11–12, 39–40, 65
  permanence of, 92–95, 102–103, 233
  and personal information, 40, 45–49,
    65–66, 233
operating systems (OS), 17, 57, 64, 66, 74,
    78, 113–114, 176–179
out-of-band verification, 62, 66, 189

P
paper records, 19, 217, 221, 233
passwords, 45, 59, 100, 168, 190–191, 207,
    215–216, 222, 225
patches
  and air gap systems, 177. *See also* air-
    gap networks
  asking questions about, 82, 111
  cloud services, 111, 114, 125
  and critical data exposure, 74–76
  internet-facing systems, 82, 216,
    232
  legacy systems, 232
  operating systems, 57, 74, 114
  servers, 82–83, 85, 108–110, 113
  software, 176. *See also* updates,
    software
PCI DSS, 197–198
permanence of online activity, 92–95,
    102–103, 233
personally identifiable information
    (PII), 3, 123, 130, 151
phishing, 29–30, 58–66, 77, 85, 153, 186,
    188–190
photos, 48, 92–95, 233
physical boundaries, 39, 133–134,
    201–206, 211–212
physical world versus virtual world,
    133–134, 201–205, 211
PINs, 191
Platform as a Service (PAAS; PAS),
    112–114, 121, 125
port scans, 178
power grid, 23–24, 67–68, 173, 219. *See also*
    infrastructure, cyberattacks on
preventing attacks versus detecting and
    controlling, 44, 163–164, 199, 234
privacy laws, 196–198. *See also* Health
    Insurance Portability and Account-
    ability Act (HIPAA)
programmable logic controller (PLC),
    54–55, 162, 167

R

ransomware attacks, 32–36, 183, 208–209, 220, 231–232
regulations, 196–198, 221. *See also* legislation
remote work, 11–12, 56–57, 132. *See also* home offices
reporting breaches, 69–70, 85, 130, 154, 164
responsibility for cybersecurity
  cloud services, 107, 110, 114–115, 125
  data centers, 108, 113, 115
  misconceptions about, 17–18, 41
  security team, 108, 136–137
  ultimate responsibility, 12, 17–20, 30, 36, 40–44, 65, 103, 108, 127–128, 131–134, 154, 183, 196, 206, 212, 222–223, 226
restaurants, skimming on tips, 187–188
risk, asking questions about, 144, 223
risk/benefit analysis. *See* functionality versus security
routers, 12, 161
Russia
  attacks for financial gain, 158, 169, 185, 211–212
  and attribution issues, 173
  elections, interference in, 193–194
  government involvement in cybercrime, 210
  and international laws, 133–134, 203, 208, 210, 212
  internet, disconnecting from, 25, 172
  power grid and infrastructure in US, access to, 23–24, 26, 158–159, 162, 169, 172
  Russian Business Network (RBN), 208
  servers located in, 203
  targets of attacks, 15
  travel to, security precautions for, 49, 66. *See also* travel, security precautions for
Russian Business Network (RBN), 208

S

Salesforce, 17, 106
Saudi Aramco, 24
Securities and Exchange Commission (SEC), 23–24
security
  asking questions about, 42, 223, 231–232
  built in, 97, 103, 132–133, 223, 227–228
  cloud services, 108–112, 115, 124, 132
  monitoring. *See* monitoring
  responsibility for. *See* responsibility for cybersecurity
  software, 55–56, 72
  team. *See* security department
security department
  cloud services, monitoring, 111–112
  communication with executives, 137–138
  contracts, responsibility for, 112, 115
  critical data and assets, responsibility for, 145–146, 154
  data center responsibilities, 108
  failure to detect attacks, 53–54, 56, 66
  financial impact of attacks, determining, 84–85, 137–138
  funding of, 108–109
  reliance on, 17, 131–132, 136, 226
  risk analysis, providing information for, 84–85, 144
security operations center (SOC), 56
servers
  asking questions about, 82, 118, 146
  cloud services, 105–106. *See also* cloud services
  dark web, 12. *See also* dark web
  data centers, 106. *See also* data centers
  internet-facing, 73–74, 81–83, 85, 106, 166–169, 193–194, 211, 215–216, 231–232
  location of, 115–118, 203
  multitiered architecture, 44
  patches, 82–83, 85, 108–110, 113
  slow performance of, 54–55, 81
  as target of cyberattack, 29, 73–77, 85, 211
service level agreements (SLAs), 109–112, 115, 125
sexting, 94
smart devices, 12, 21, 23, 41, 78. *See also* mobile devices
Snapchat, 92
Sneakers (film), 214
social media, 46, 48, 60, 63, 65, 90, 92
social security numbers, 3–4, 12–13, 75, 183–184, 213–214
software, 38, 40, 54–56, 72, 74, 76, 82–83, 85

Software as a Service (SAAS; SAS), 112–114, 125
spoofing, 28, 59, 189–191
Stuxnet, 54–55, 162–163
Supervisory Control and Data Acquisition systems (SCADA), 165
supply chain, 130–131, 154, 160–161
Sytex Group, Inc., The (TSGI), 9

T
T-Mobile breach, 41
Target breach, 13, 42–43
targets of cybersecurity attacks
    all individuals and entities, 12, 15–16, 26–30, 40–41, 65, 73–78, 84–85, 127–131, 154, 183, 223–226
    critical assets, 134–138, 155
    critical data. *See* critical data
    and customers, 129
    factors for becoming target, 128–130, 154
    financial gain as motive, 169–170
    financial sector. *See* finance sector
    healthcare sector. *See* healthcare sector
    identifying, 29–30
    inevitability of attack, 140
    infrastructure. *See* infrastructure, cyberattacks on
    intellectual property. *See* intellectual property
    and location of business, 128–129, 154
    misconceptions about, 15
    and name of business, 128, 154
    and online activity, 11–12, 40, 65
    power grid, 23–24, 67–68, 173, 219
    predicting, 47
    servers. *See* servers
    supply chain, 130–131, 154

text messages, 28, 39, 59–60, 62, 66, 94–96, 102–103, 186–188, 190, 233
threat hunting, 140, 148
TikTok, 90
travel, security precautions for, 49–50, 66, 151
TSGI (The Sytex Group, Inc.), 9
tunneling, 171
two-factor authentication (2FA), 57–59, 65–66, 100, 103, 191, 207, 222, 225

U
United Kingdom (UK)
    and cyber war, 158, 170
    WannaCry attack, 32, 34, 182–183, 208–209
updates, software, 74, 76, 82–83, 85
USB drives, 20–21, 222
user IDs, 106, 168, 207, 215–216

V
verification, out-of-band, 62, 66, 189, 199
virtualization, 113
VPN (virtual private network), 65, 107

W
WannaCry, 32, 34, 182–183, 208–209
WarGames (film), 214–215
web browsers, 61, 64, 66, 228
web links, 30, 32–33, 63–64, 73, 77, 82–85, 96, 152–153, 183, 223, 226, 228–230
websites, 229–230
WhatsApp, 90
Windows, 64, 66, 78, 176, 178–179, 228–229
wire transfers, fraudulent, 58–60, 62
working from home. *See* home offices

Z
Zoom, 133, 227

About the Author

Photo by Shelby Au

DR. ERIC COLE is an industry-recognized security expert with more than twenty years of experience in information technology. His work focuses on helping customers identify specific areas of potential compromise and then building dynamic defense solutions that protect their organizations from advanced threats.

Dr. Cole is the author of one of the best-selling courses on cybersecurity for over fifteen years and has trained over 65,000 people at various events around the world on the importance of cybersecurity and the impact it can have on any entity. Dr. Cole is on a mission to make cyberspace a safe place to live, work, and raise a family.

His other books include:

Advanced Persistent Threat: Understanding the Danger and How to Protect Your Organization

Hackers Beware: The Ultimate Guide to Network Security

Hiding in Plain Sight: Steganography and the Art of Covert Communication

Network Security Bible

Insider Threat: Protecting the Enterprise from Sabotage, Spying, and Theft

Online Danger: How to Protect Yourself and Your Loved Ones from the Evil Side of the Internet

Dr. Cole holds a master's degree in computer science from New York Institute of Technology and a doctorate from Pace University, with a concentration in information security. He was a member of the commission on cybersecurity for the forty-fourth president and is a member of several executive advisory boards. He was inducted into the 2014 Infosecurity Hall of Fame.

Dr. Cole is founder and CEO of Secure Anchor Consulting, which provides state-of-the-art security services and expert witness work. He also served as CTO of McAfee, chief scientist for Lockheed Martin, and worked at the CIA for more than eight years.